PHYSICAL GEOGRAPHY LABORATORY MANUAL

KAREN A. LEMKE
University of Wisconsin—Stevens Point

MICHAEL E. RITTER
University of Wisconsin—Stevens Point

N.C. HEYWOOD
University of Wisconsin—Stevens Point

McGraw-Hill Higher Education

Boston   Burr Ridge, IL   Dubuque, IA   New York   San Francisco   St. Louis
Bangkok   Bogotá   Caracas   Kuala Lumpur   Lisbon   London   Madrid   Mexico City
Milan   Montreal   New Delhi   Santiago   Seoul   Singapore   Sydney   Taipei   Toronto

PHYSICAL GEOGRAPHY LABORATORY MANUAL

Published by McGraw-Hill, a business unit of The McGraw-Hill Companies, Inc., 1221 Avenue of the Americas, New York, NY 10020. Copyright © 2009 by The McGraw-Hill Companies, Inc. All rights reserved. No part of this publication may be reproduced or distributed in any form or by any means, or stored in a database or retrieval system, without the prior written consent of The McGraw-Hill Companies, Inc., including, but not limited to, in any network or other electronic storage or transmission, or broadcast for distance learning.

Some ancillaries, including electronic and print components, may not be available to customers outside the United States.

 This book is printed on recycled, acid-free paper containing 10% postconsumer waste.

Printed in the United States

5 6 7 8 9 0 DOW/DOW 16 15

ISBN 978–0–07–727603–4
MHID 0–07–727603–5

Publisher: *Thomas Timp*
Executive Editor: *Margaret J. Kemp*
Senior Developmental Editor: *Joan M. Weber*
Senior Marketing Manager: *Lisa Nicks*
Project Manager: *Joyce Watters*
Senior Production Supervisor: *Kara Kudronowicz*
Lead Media Project Manager: *Judi David*
Senior Designer: *David W. Hash*
 (USE) Cover Image: *©Eyewire: Seasons EP083/GETTY IMAGES (RF)*
Senior Photo Research Coordinator: *John C. Leland*
Photo Research: *Jo Hebert*
Compositor: *Electronic Publishing Services Inc., NYC*
Typeface: 10/12 *Times*
Printer: RR Donnelley

www.mhhe.com

TABLE OF CONTENTS

Preface   v

About the Authors   vii

EXERCISE 1.   Earth-Sun Geometry and Insolation   1

EXERCISE 2.   Radiation and Energy Balance at the Earth's Surface   11

EXERCISE 3.   Atmospheric Temperature   19

EXERCISE 4.   Atmospheric Pressure, Circulation, and Wind   31

EXERCISE 5.   Water in the Atmosphere   41

EXERCISE 6.   Lapse Rates, Adiabatic Processes, and Cloud Development   49

EXERCISE 7.   Midlatitude Weather and Weather Map Interpretation   59

EXERCISE 8.   Climate Classification and Regional Climates   71

EXERCISE 9.   Soil Moisture Budgets   93

EXERCISE 10.   Analysis of Soil Moisture Properties   111

EXERCISE 11.   Climate, Net Primary Production, and Decomposition   125

EXERCISE 12.   Vegetation Form and Range   141

EXERCISE 13.   Bioclimatic Transects   149

EXERCISE 14.   Coincident Climates, Vegetation, and Soils   163

EXERCISE 15.   Hawai'i Rainforest Regeneration   175

EXERCISE 16.   Introduction to Topographic Maps   185

EXERCISE 17.   Igneous Landforms   205

EXERCISE 18.   Drainage Basin Analysis   221

EXERCISE 19.   Fluvial Landforms   245

EXERCISE 20.   Glacial Landforms   255

EXERCISE 21.   Coastal Landforms   271

APPENDIX A.   Units of Measure and Conversions   287

APPENDIX B.   Drawing Isolines   290

APPENDIX C.   Constructing Profiles   292

APPENDIX D.   Using Pocket Stereoscopes   294

APPENDIX E.   Exercise Maps and Photos   297

APPENDIX F.   Thematic World Maps   324

Index   329

PREFACE

We wrote this laboratory manual because most currently available laboratory manuals do not provide equal coverage of the four spheres of the environment—the atmosphere, biosphere, hydrosphere, and lithosphere. They focus primarily on the atmosphere and lithosphere, so we've written a laboratory manual that we feel provides more balanced coverage. There are eight exercises that address topics related to the atmosphere, seven exercises that address topics related to the biosphere (including soils), and six exercises that address topics related to the lithosphere, for a total of 21 exercises. Of these 21 exercises, five address topics related to the hydrosphere and overlap with topics in the other three spheres.

Most laboratory manuals do not require as much critical thinking as we wanted our students to engage in. Although many of these manuals ask students to do things, if students are not required to follow up their activities with thoughtful questions, what they've done may not be meaningful or memorable. Active learning doesn't just mean "doing things"; it includes actively thinking about what we're doing or what we've done, and what the results of our activities show us. Thus, we've tried to provide more thought-provoking questions and activities, and we've tried to follow up on activities by asking questions that require students to go back and look at what they've done and think about what it means.

Finally, we wanted to ensure that our students receive a scientifically rigorous experience in physical geography.

FEATURES

All exercises were written independently—it is not necessary to do exercise 1 before doing exercise 2. As a result, these exercises can be done in the order most appropriate for your class. This manual was written independent of any specific text, and should work with most of the currently available textbooks on physical geography.

Each exercise contains a brief statement regarding the purpose of the exercise. This is followed by a list of specific learning objectives. These learning objectives are testable items that will allow instructors to objectively assess what the students have learned. The questions and activities that follow help students achieve the stated learning objectives. Some exercises are divided into more than one part to allow flexibility in what topics are covered during a laboratory period, and if not all parts are completed some of the learning objectives may not be accomplished. Thus, the learning objectives address what students should be able to do if they complete the entire exercise. We do not always do all the parts of each exercise each semester, and we let our students know which learning objectives they are responsible for.

Each exercise contains an introduction that addresses topics specifically covered in the exercise activities. The introduction is not meant to replace a full treatment of the topic like that found in a textbook; however, since not all textbooks contain equal coverage of the topics in the laboratory manual, we included enough detail in the introduction to ensure that students have enough background in the material to complete the exercise. Information on topics not specifically addressed in the exercise is not included in the introduction. The length of the exercise introductions varies depending on the degree to which the topic is usually covered in a typical textbook. For example, most textbooks provide ample coverage of adiabatic lapse rates, and thus the introduction to this exercise is relatively brief. Not all typical textbooks cover soil water budgets, and as a result, the introduction to this exercise is lengthier.

At the end of the introduction, there is a list of important terms, phrases, and concepts. These include all the bold-faced items defined in the introduction. At a minimum, students should be familiar with these phrases and concepts, knowing their definition, why they're important, and when and how to use this information for solving geographical problems.

TEACHING AND LEARNING SUPPLEMENTS

McGraw-Hill offers various tools and technology products to support *Physical Geography Laboratory Manual*. Students can order supplemental study materials by contacting their local bookstore or by calling 800-262-4729. Instructors can obtain teaching aids by calling the Customer Service Department at 800-338-3987, visiting the McGraw-Hill website at www.mhhe.com, or by contacting their local McGraw-Hill sales representative.

A password-protected website can be found at www.mhhe.com/lemke1e. This helpful resource includes an *Instructor's Manual* and an extensive array of teaching and learning tools. Visit this text-specific website today!

ACKNOWLEDGMENTS

The authors would like to thank their families for their support over the years as we've worked on this laboratory manual. We also thank our colleagues at the University of Wisconsin—Stevens Point for help and support, as well as reviewers who provided useful comments and suggestions. Reviewers who provided many valuable suggestions include

David M. Cairns, *Texas A&M University*

Christopher H. Exline, *University of Nevada—Reno*

Doug Goodin, *Kansas State University*

Linda Lea Jones, *Texas Tech University*

Hsiang-te Kung, *University of Memphis*

Miles R. Roberts, *California State University—Sacramento*

Robert Rohli, *Louisiana State University*

Dean Wilder, *University of Wisconsin—La Crosse*

The authors wish to express special thanks to McGraw-Hill for editorial support through Marge Kemp and Joan Weber; the marketing expertise of Lisa Nicks; and the production team led by Joyce Watters, David Hash, John Leland, Kara Kudronowicz, and Sandy Schnee.

And last, we would like to thank our students, because without them, we never would have written this.

Karen A. Lemke
Michael E. Ritter
N.C. Heywood

ABOUT THE AUTHORS

KAREN A. LEMKE Karen A. Lemke is a professor of geography and geology at the University of Wisconsin—Stevens Point (UWSP). She received her bachelor's degree in 1981 at Bucknell University with majors in geography and German. She received her Ph.D. in 1988 in geography from the University of Iowa. She currently teaches introductory physical geography, geomorphology, glacial geology, and applied statistics in geography. In 1997 she won a teaching excellence award from UWSP. Her research interests are in fluvial geomorphology and the scholarship of teaching and learning.

MICHAEL E. RITTER Michael E. Ritter received his undergraduate degree in geography from Western Illinois University. He went on to receive a master's degree and Ph.D. in geography from Indiana University. Michael is a professor of geography in the Geography & Geology Department at the University of Wisconsin—Stevens Point. His primary teaching responsibilities include physical geography and climatology. Michael wrote the first book about using the Internet for earth science education and research, a book titled *Earth Online*. He has also written one of the first interactive online textbooks in physical geography, titled *The Physical Environment*. Michael has served as a media author, contributor, and consultant for various publishers. His research interests focus on the use of interactive multimedia learning technologies in geoscience.

N.C. HEYWOOD After childhood in the northeastern United States, international experience as a U.S. Navy navigator, and a cartography internship with the National Geographic Society, N.C. Heywood completed B.A. (SUNY—Plattsburgh, 1982), M.A. (Georgia, 1984), and Ph.D. (Colorado, 1989) degrees—all in geography. After joining the Geography & Geology faculty at the University of Wisconsin—Stevens Point in 1989, N.C. continues as a physical geographer specializing in environmental hazards, biogeography, field methods, and career development. N.C. received a teaching excellence award in 1998 and is currently working on research assisting the Wisconsin Department of Natural Resources Forestry Division, the U.S. Department of Agriculture Forest Service, and the U.S. Department of Interior National Park Service.

EXERCISE 2 † RADIATION AND ENERGY BALANCE AT THE EARTH'S SURFACE

PURPOSE

The purpose of this exercise is to develop an understanding of the spatial and temporal controls on the radiation and energy balance at the earth's surface.

LEARNING OBJECTIVES

By the end of this exercise you should be able to

- calculate and interpret the radiation balance at a site;
- explain how the radiation balance changes with season; and
- explain how the partitioning of energy use affects temperature and humidity.

INTRODUCTION

Most environmental processes active at the earth's surface and in the earth's atmosphere involve the exchange of energy in one form or another. Energy is exchanged between the earth's surface and the atmosphere via radiation, conduction, and convection. All bodies that have a temperature emit radiation. The laws of physics tell us that:

1. The intensity of radiation increases with increasing temperature. That is, very hot bodies like the sun (6,270°C or 11,318°F at its surface) emit very intense radiation. The earth by comparison emits much less energy, as its globally averaged temperature is only approximately 15°C (59°F).
2. The wavelength of emission decreases with increasing temperature; or the hotter the body, the shorter the wavelength of emission. Therefore the sun, being an extremely hot body, emits primarily shortwave radiation while the earth emits longwave radiation.

Although energy is unequally distributed across the earth, there is still a *global balance* between the total incoming and total outgoing radiation. If this were not true, the earth would either continue to heat up or cool down.

On a local scale, however, there may be differences between incoming and outgoing radiation, depending on: (1) the time of day, (2) the time of year, and (3) the surface characteristics. The albedo and the heat capacity of different surfaces are important for determining the local heat balance.

Applying the principles of an energy budget at the global scale to a local area, it is possible to calculate a specific energy budget for different localities. Such detailed studies provide an indication of the net radiation of an area and may have some use in planning and design of buildings using solar energy or energy conservation techniques.

THE RADIATION BALANCE

The **radiation balance** is an expression of the inputs and outputs of radiant energy at the earth's surface. Shortwave solar radiation as well as energy emitted by the atmosphere (longwave radiation) is received and absorbed at the earth's surface. Radiant energy is also emitted and reflected by the earth. The difference between these incoming and outgoing radiation components is net radiation:

$$Q^* = K\downarrow - K\uparrow + L\downarrow - L\uparrow \qquad (2.1)$$

where:

Q^* = net radiation

$K\downarrow$ = incoming shortwave radiation

$K\uparrow$ = outgoing (reflected) shortwave radiation

$L\downarrow$ = incoming longwave radiation; longwave radiation emitted by the atmosphere and directed toward the ground

$L\uparrow$ = outgoing longwave radiation; longwave radiation emitted by the earth's surface and directed toward the sky

Equation 2.1 can be rewritten to show more detail regarding the shortwave portion of the radiation balance:

$$Q^* = [(S+D) - ((S+D)a)] + (L\downarrow - L\uparrow) \qquad (2.2)$$

where:

Q^* = net radiation

S = direct solar radiation

D = diffuse solar radiation

a = surface albedo

$L\downarrow$ = incoming longwave radiation

$L\uparrow$ = outgoing longwave radiation

Shortwave Radiation

Insolation, or incoming solar radiation, is composed of both **direct beam shortwave radiation** and **diffuse (scattered) shortwave radiation.** The amount of insolation reaching the surface varies from hour to hour and day to day due to changes in the sun angle, the day length period, the condition of the atmosphere, and the atmospheric path length. The proportion of total incoming solar radiation that is diffuse radiation (as opposed to direct beam radiation) increases as the cloud cover and water vapor content of the atmosphere increases. Thus humid climates are typically more cloudy than dry climates, resulting in higher proportions of diffuse radiation. Seasonal variations in insolation are largely explained by changing sun angles.

The quantity of shortwave radiation reflected from the earth's surface, **outgoing shortwave radiation,** is determined by surface characteristics. The albedo of the surface is a measure of its reflectivity. The **albedo** (a) is the proportion of insolation (S+D) that is reflected by a surface. Albedos range from 0 (or 0%, i.e., no reflection) to 1 (or 100%, i.e., total reflection). Surfaces that absorb a lot of incoming energy have a low albedo, such as a dark forested area (a = 0.20); therefore, little of the energy striking the surface is reflected away (100% × .20 = 20%). Some surfaces, such as fresh flat snow surfaces, have a high albedo and hence reflect a large proportion of insolation back toward the sky. Thus shortwave radiation lost from the earth's surface is simply the amount of insolation that is reflected back to space, and can be calculated as the product of the surface albedo times the incoming shortwave radiation:

$$K\uparrow = ((S+D)a) \quad (2.3)$$

where:

$K\uparrow$ = the amount of reflected shortwave radiation

S = direct solar radiation

D = diffuse solar radiation

a = albedo

Table 2.1 lists some typical albedos.

Longwave Radiation

The earth and its atmosphere radiate energy. Because the temperature of the air and the surface are so much cooler than the sun, these bodies emit primarily longwave radiation. Recall from the introduction that the amount of energy emitted by a body is proportional to its temperature. Thus **outgoing longwave radiation** (radiation emitted by the earth) is high for places that absorb much solar radiation, resulting in high surface temperatures. Radiation emitted by the atmosphere (**incoming longwave radiation**) depends on the temperature of the atmosphere. In general, temperatures tend to increase with increasing insolation and decreasing latitude (high sun angles closer to the equator). Temperatures tend to be lower

TABLE 2.1 TYPICAL ALBEDOS

Fresh snow	0.75–0.95	(75%–95%)
Old snow	0.40–0.70	(40%–70%)
Soil, dark	0.05–0.15	(5%–15%)
Asphalt	0.05–0.10	(5%–10%)
Deciduous forest	0.10–0.20	(10%–20%)
Cumulus cloud	0.75–0.90	(75%–90%)
Grass	0.15–0.25	(15%–25%)
Sand	0.20–0.40	(20%–40%)

along coastlines and at high elevation in mountain regions. In general, the earth's surface tends to be warmer than the atmosphere, resulting in more outgoing longwave radiation than incoming longwave radiation.

Net Radiation

Net radiation (Q^*) at the earth's surface is a function of those factors affecting incoming and outgoing radiation. Net radiation is equal to the total energy gained or lost at the earth's surface and may be represented as the balance between radiation "in" and radiation "out." Radiation gained includes incoming shortwave radiation and longwave radiation emitted from the atmosphere toward the earth. Radiation lost, on the other hand, includes shortwave radiation reflected from the earth back out to space, and longwave radiation emitted from the earth's surface out to space. Net radiation then is the amount of energy available at the earth's surface to do work in the environment.

In tropical and subtropical latitudes, net radiation is positive throughout the year because of the high sun angles (much insolation) and high water vapor content or cloud coverage (large incoming longwave flux or transfer). In high latitudes where sun angles are low in winter and snow albedos high, the annual net radiation is negative. This means that low latitudes are gaining radiant energy while high latitudes are losing energy. Therefore, heat must be transported from energy surplus regions (low latitudes) to energy deficient regions (high latitudes) by the circulation of the atmosphere and of the oceans in order to keep the Poles from continually cooling and the tropics from overheating.

On a daily basis for a particular location, net radiation tends to be positive during most of the daylight hours. During this period, insolation is high and much radiant energy is gained at the surface. During the night, incoming solar radiation is nonexistent. Therefore, the radiant transfers near the surface are longwave radiation only. Under normal nighttime circumstances, the earth's surface remains warmer than the atmosphere. Consequently, the earth loses more longwave radiation than it gains from incoming atmospheric radiation. Under these conditions, net radiation would be a negative value.

THE ENERGY BALANCE

The **energy balance** is an accounting of the major utilizations of energy, mainly radiant energy from the sun. Energy is used for heating the atmosphere, heating the subsurface, and evaporating water. These utilizations require a transfer (flux) of energy from a source to a receptor. **Energy fluxes** always occur from sources with a high energy content to receptors with a lower energy content. Energy is transferred from sources to receptors either by conduction, convection, or radiation. **Radiation** is the transfer of energy by electromagnetic waves and requires no intervening medium through which to pass. **Conduction** is the transfer of energy by molecular collisions. **Convection** is the transfer of energy by the circulation (movement) of a heated mass, such as a liquid or a gas. Conduction and convection are referred to as **non-radiative fluxes.** As energy is transferred, the source cools down as it loses energy, while the receptor warms up as it gains energy.

It is important to account for energy flows at the surface of the earth, thus the surface of the earth can be thought of as the interface between the atmosphere above and the solid subsurface below. (If the surface were water, lake, pond, or ocean, the subsurface would obviously be a liquid not a solid). Figure 2.1 illustrates one way in which energy flows can be diagrammed using a particular convention for the plus and minus signs. Plus signs indicate non-radiative transfers of energy away from the surface. Minus signs indicate non-radiative transfers of energy toward the surface. Using the same convention for plus and minus signs, the energy balance can be expressed as an equation:

$$Q^* = LE + H + G \qquad (2.4)$$

where:

- Q^* = net radiation from equation (2.1) or (2.2).

- LE = **latent heat flux,** the energy gained or lost by water during evaporation, condensation or freezing. When LE is positive, energy is gained by water and evaporation takes place. This usually results in water vapor being convected into the air thus removing energy from the evaporating surface. A negative latent heat flux occurs during condensation when latent heat is released from the water vapor to the surrounding environment.

- H = **sensible heat flux,** the gain or loss of energy by the atmosphere. Sensible heat transfer occurs by conduction between the surface and the air lying directly above the surface, and then by convection to greater heights. It is called sensible heat flux because the result of this transfer can be felt as an increase or decrease in air temperature. When H is positive, heat is transferred from the surface into the air above. When H is negative, heat is transferred from the air to the surface.

- G = **ground heat flux,** the energy gained or lost by the subsurface. This kind of heat transfer occurs by conduction. When G is positive, heat is transferred from the surface downward to the subsurface. When G is negative, heat is transferred from the subsurface upward toward the surface.

FIGURE 2.1 ENERGY BALANCE FLUXES AT THE SURFACE OF THE EARTH

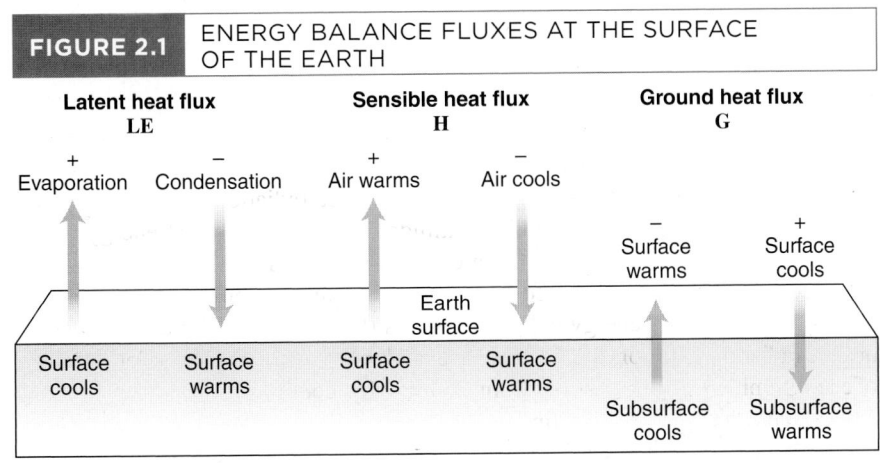

The respective energy utilizations depend on characteristics of the place under consideration. For instance, at a desert site where there is little water available for evaporation, most of the net radiation is used for H and G. Most deserts have high temperatures, in part, as a result of large positive sensible heat transfer (H) into the air (though there are exceptions). Regions of low sensible heat transfer (H) experience cool temperatures. At sites where water is abundant due to precipitation or location near a large body of water, however, most net radiation is utilized for evaporation (LE), not H. Over moist surfaces approximately 70% of net radiation is used for evaporation, the remaining is used mostly for sensible heat flux. These places tend to be very humid as much water vapor and latent heat are transferred into the air.

IMPORTANT TERMS, PHRASES, AND CONCEPTS

radiation balance
insolation
direct beam shortwave radiation
diffuse (scattered) shortwave radiation
outgoing shortwave radiation
albedo
outgoing longwave radiation
incoming longwave radiation
net radiation

energy fluxes
radiation
conduction
convection
non-radiative fluxes
latent heat flux
sensible heat flux
ground heat flux
energy balance

Name: _____   Section: _____

PART 1 † GEOGRAPHICAL VARIATIONS IN THE RADIATION BALANCE

Use the data in Table 2.2 to answer questions in Part 1.

TABLE 2.2 RADIATION BALANCE DATA

Bondville, Illinois 40.05N; 88.37W 213 meters		Desert Rock, Nevada 36.63N; 116.02W 1007 meters	
Month 1	Month 2	Month 1	Month 2
S = 33.5 W/m²	S = 168.7 W/m²	S = 73.5 W/m²	S = 266.7 W/m²
D = 37.7 W/m²	D = 107.9 W/m²	D = 37.9 W/m²	D = 67.9 W/m²
a = 0.55	a = 0.19	a = 0.20	a = 0.21
L↓ = 277.7 W/m²	L↓ = 385.2 W/m²	L↓ = 286.6 W/m²	L↓ = 372.7 W/m²
L↑ = 305.7 W/m²	L↑ = 443.1 W/m²	L↑ = 353.0 W/m²	L↑ = 518.9 W/m²
RH* = 83.6 %	RH = 76.9%	RH = 61.7%	RH = 19.9%

* RH = Relative humidity

Source: Data from NOAA/SURFRAD Network

1. Calculate total incoming solar radiation (S+D) for each location during each month.

   Bondville   Desert Rock
   Month 1 _____   Month 1 _____
   Month 2 _____   Month 2 _____

2. Calculate the amount of reflected solar radiation (K↑) (equation 2.3), for each month.

   Bondville   Desert Rock
   Month 1 _____   Month 1 _____
   Month 2 _____   Month 2 _____

3. Calculate the net radiation (Q*) for each month (equation 2.2).

   Bondville   Desert Rock
   Month 1 _____   Month 1 _____
   Month 2 _____   Month 2 _____

4. Use the values of the albedo and Table 2.1 to determine the type of surface over which the data was collected.

   Bondville   Desert Rock
   Month 1 _____   Month 1 _____
   Month 2 _____   Month 2 _____

15

5. a. Carefully examine the values of L↓ and L↑ for Month 1. Which is warmer, the surface or the air during Month 1?

   Bondville _____   Desert Rock _____

   b. Carefully examine the values of L↓ and L↑ for Month 2. Which is warmer, the surface or the air during Month 2?

   Bondville _____   Desert Rock _____

6. a. Based on the values of L↓, which month had a higher air temperature, Month 1 or Month 2?

   Bondville _____   Desert Rock _____

   b. Based on the values of L↑, which month had a higher surface temperature, Month 1 or Month 2?

   Bondville _____   Desert Rock _____

7. a. Using your answers to questions 1 through 6, suggest the season of year, summer or winter, for each month.

   Month 1 _____   Month 2 _____

   b. Explain how you came to your conclusion.

8. a. Which location had the highest surface temperatures in both seasons?

   b. Why?

9. Although Desert Rock is located in the Nevada desert, why does it have a lower value of L↓ during Month 2 than Bondville? (Hint: Review the data table and consider the factors that influence air temperature.)

10. a. Calculate the percent of diffuse radiation for each month as:

    $$[D/(S + D)] \times 100\%$$

    Bondville   Desert Rock

    Month 1 _____   Month 1 _____

    Month 2 _____   Month 2 _____

    b. What likely accounts for the differences in percent diffuse radiation during these two months at each site?

    c. Based on your answers, speculate on why these differences exist. (Hint: Consider location and site relative humidity, a measure of the water vapor content of the air.)

Name: _____   Section: _____

PART 2 · GEOGRAPHICAL VARIATIONS IN THE ENERGY BALANCE

Table 2.3 provides representative energy balance data for three sites with differing climatic characteristics. Use this data to answer the questions below.

TABLE 2.3 ANNUAL AVERAGE ENERGY BALANCE DATA

Site A	Site B	Site C
50°N, 100°W	20°N, 22°E	0°, 70°W
H = 13 W/m²	H = 55 W/m²	H = 23 W/m²
LE = 60 W/m²	LE = 15 W/m²	LE = 65 W/m²
Q* = 75 W/m²	Q* = 90 W/m²	Q* = 90 W/m²

1. What percent of net radiation is sensible heat for each of the three sites?

    Site A = _____   Site B = _____   Site C = _____

2. What percent of net radiation is latent heat for each of the three sites?

    Site A = _____   Site B = _____   Site C = _____

3. a. Which site probably had the warmest air temperature? _____

    b. What evidence did you use to support your answer?

4. a. Which site probably had the highest rate of evaporation? _____

    b. What evidence did you use to support your answer?

5. Describe the climate conditions (warm/cold—humid/dry) that are represented by:

    Site A: _____

    Site B: _____

    Site C: _____

6. What evidence did you use to support your answers for question 5?

7. Match the site to the climate type by plotting the latitude and longitude of each site on a world climate map.

   Site A   Tropical rain forest climate

   Site B   Humid continental climate

   Site C   Subtropical desert climate

8. Describe the climatic conditions for each of the climates as you did in question 5, e.g., warm/cold—humid/dry. (Hint: You may want to consult your textbook.)

9. Sites B and C have similar available energy (net radiation) but very different utilization for latent heat. Explain why.

10. Summarize the relationship between energy utilization, temperature, and humidity in each of the three climates.

EXERCISE 6 † LAPSE RATES, ADIABATIC PROCESSES, AND CLOUD DEVELOPMENT

PURPOSE

The purpose of this exercise is to learn about adiabatic processes and conditions of atmospheric stability or instability, since these conditions may lead to precipitation.

LEARNING OBJECTIVES

By the end of this exercise you should be able to

- determine the stability of the air;
- explain how air density changes in rising parcels of air;
- explain how temperature and humidity changes in rising and sinking air; and
- determine the condensation level and influence of stability on cloud development.

INTRODUCTION

Precipitation and moisture in the atmosphere are important determinants of local and global weather and climate patterns. Water vapor in the atmosphere also plays an important role in the transfer of heat energy around the earth. To understand the patterns of rainfall around the world, including the amount and timing of precipitation, it is necessary to understand the various mechanisms that cause air to cool beyond the point of saturation so that condensation and cloud development occur. Without a knowledge of these mechanisms, it is difficult to predict the weather accurately or understand why seasonal patterns of rainfall exist in many places.

Lapse Rates

One way to cool air is to have the air rise. If a parcel of air is forced upward, for whatever reason, there may be a significant change in the weather conditions. The altitude to which air rises determines the type of clouds that develop, the amount of precipitation, and storm activity. The higher the air rises, the more cooling, condensation, and continued precipitation will take place. It is important, therefore, to understand the factors that may affect the extent to which air will rise.

The **normal lapse rate** is the rate of change in temperature with altitude in the troposphere. The normal lapse rate averages 0.65°C/100 meters, but the *actual* rate of change for different locations, or for a specific location at different times, may vary considerably. The **environmental lapse rate** (ELR) of temperature is the actual rate of change on a given day for a particular location. The environmental lapse rate can be larger or smaller than the normal lapse rate of temperature. Vertical mixing or horizontal winds at the earth's surface and aloft cause the rate of change to vary between different altitudes, and at times may cause temperatures to increase with an increase in altitude, resulting in a **temperature inversion.** An **isothermal layer** occurs when there is no change in temperature as altitude changes. Because the ELR is not constant, and it varies from place to place and day to day, the only way to determine the ELR is to take temperature measurements at various heights throughout the troposphere. This is done with weather balloons or radiosondes that broadcast the information back to weather stations.

Adiabatic Processes

If air moves vertically through the atmosphere, it will undergo a variety of physical changes. When a parcel of air rises, the surrounding atmospheric pressure decreases. As a result, the parcel expands in size (parcel volume increases). Because the number of molecules in the parcel does not change as the volume increases, the density of molecules decreases and the air becomes more buoyant. Some of the air's internal energy is used to expand the parcel, which causes the air temperature to decrease. Likewise, as air sinks, the surrounding atmospheric pressure increases, forcing the air to compress (parcel volume decreases). This causes the air parcel to warm. As the volume of the parcel decreases, its density increases. Temperature change brought on by expansion or compression of a parcel of air is called an **adiabatic temperature change.** The rate of temperature change resulting from expansion or compression is an **adiabatic lapse rate** of temperature. The rate of heating or cooling of vertically moving unsaturated air is fairly constant and is referred to as the **dry adiabatic lapse rate** (DALR) and equals 1°C/100 meters.

If air is saturated, it still cools or warms due to adiabatic processes but there is also a new source of heat that must be considered, the **latent heat of vaporization** (energy stored or released during condensation or evaporation). As air rises and cools, its ability to keep water in its vapor state decreases. If the air cools enough, it will eventually reach its **dew point,** the temperature at which air becomes saturated. The **condensation level** is the elevation at which rising air becomes saturated. If the air continues to rise and cool, condensation and cloud growth occur. As the air continues to rise and cool beyond this level, it stays saturated, which means that condensation is a continual but gradual process. When water vapor condenses, heat is released. The release of heat to the surrounding air offsets the rate of cooling due to expansion. Likewise, saturated air that sinks will warm at a slower rate than unsaturated air because heat is absorbed (or taken away) during evaporation, and this offsets the rate of warming due to compression. Hence, the rate of heating or cooling of vertically moving saturated air is less than that of unsaturated air, and is referred to as the **saturated adiabatic lapse rate** (SALR). The saturated adiabatic lapse rate is not constant and depends on the original temperature and moisture content of the air, but an average value of 0.6°C/100 meters is commonly used.

Cloud Development

A parcel of air will rise through the atmosphere on its own as long as it is warmer, and less dense, than the surrounding environmental air. Once the temperature of a parcel of air is the same as the surrounding air, it will no longer rise under its own "power" and atmospheric conditions are said to be **stable.** Air has a tendency to "stand still" when it is stable. That is, it will not move vertically from its position under its own power. It is during periods of strong atmospheric stability that pollutant concentrations build up to hazardous levels in urban areas. Vertical mixing of the air is required to "ventilate" pollutants upward and away from the surface. When the air has little impetus to move upward and cool, condensation of water vapor is low. Therefore, precipitation is unlikely during stable atmospheric situations. When a parcel of air continues to rise through the atmosphere and remains warmer than the surrounding air, conditions are said to be **unstable.** Unstable conditions promote cloud development, and as a result, precipitation is more likely.

Cloud development may result from convection of warm moist air. This happens when the earth's surface is heated, especially on sunny afternoons. Energy is transferred from the surface to the air above, causing the air to warm and rise. Convectional uplift, with subsequent cloud development, is especially important along coasts where moist air masses move onshore from the ocean. During the summer, rapid heating of the ground surface compared to the ocean results in air over the land being much warmer than air over the ocean. Once heated, the warm air becomes unstable and rises, leading to the formation of cumulus clouds and possible scattered showers (convectional precipitation). Thermal convection along coasts rarely results in widespread rains.

Cloud development and orographic precipitation may result when air is forced to rise over mountains. Rising air on the windward side of mountains cools adiabatically, and if it rises beyond the condensation level, clouds will develop and precipitation often occurs. This produces wet conditions on the windward side of mountains, particularly if the mountains are in the path of prevailing winds coming off the ocean. On the leeward side, air warms adiabatically as it sinks, resulting in evaporation and the formation of rainshadow deserts.

IMPORTANT TERMS, PHRASES, AND CONCEPTS

normal lapse rate

environmental lapse rate (ELR)

temperature inversion

isothermal layer

adiabatic temperature change (adiabatic process)

adiabatic lapse rate

dry adiabatic lapse rate (DALR)

latent heat of vaporization

dew point temperature

condensation level

saturated adiabatic lapse rate (SALR)

stable atmospheric conditions

unstable atmospheric conditions

Name: _____ Section: _____

PART 1 † LAPSE RATES AND STABILITY

1. Table 6.1 below gives some information about parcels of air at five weather stations. Use the saturation curve in Figure 6.1 and your knowledge of adiabatic processes to fill in the rest of the table, assuming that the parcels of air are forced to rise.

TABLE 6.1 RISING AIR PARCELS AT FIVE WEATHER STATIONS

Weather Station	A	B	C	D	E
Air temperature at ground level (°C)	13	33	23	32	30
Water vapor content of the air (g/kg)	5	10			17
Dew point temperature (°C)			15	18	22
Condensation level (m)					

FIGURE 6.1 SATURATION CURVE

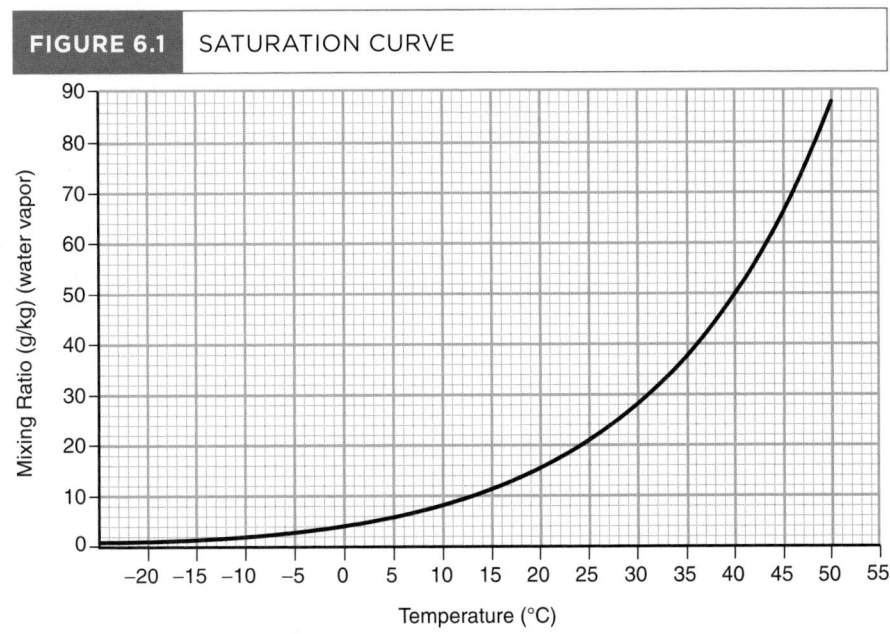

51

2. Table 6.2 provides information on the environmental lapse rate (ELR) from a weather balloon. Table 6.2 also provides information on the air temperature and dew point for two air parcels at ground level (0 meters).

   a. Draw a graph of the environmental lapse rate (ELR) data presented in Table 6.2 on the graph provided (Figure 6.2) and connect the points with straight line segments.

   b. Draw a graph that represents the adiabatic temperature change that results when parcel A is forced to rise from the surface to an elevation of 1200 meters. Do this on the same graph (Figure 6.2) as the ELR for part (a).

   c. Draw a graph that represents the adiabatic temperature change that results when parcel B is forced to rise from the surface to an elevation of 1200 meters. This should be done on the same graph (Figure 6.2) as the ELR for part (a).

   d. Label the portion of the graphs for parcels A and B (Figure 6.2) where the air is cooling at the dry and saturated adiabatic rates.

   e. Label the level at which condensation occurs and clouds develop.

   f. Decide at what elevations the parcels of air are stable or unstable and label the graph (Figure 6.2) clearly.

3. a. Between what altitudes does a temperature inversion occur?

   b. Label the graph (Figure 6.2) where the inversion occurs.

TABLE 6.2 ENVIRONMENTAL LAPSE RATE DATA AND RISING PARCELS OF AIR

Environmental Lapse Rate		Parcel A — Dew point temp. = 14°C		Parcel B — Dew point temp. = 16°C	
Altitude (m)	Air Temp. (°C)	Altitude (m)	Air Temp. (°C)	Altitude (m)	Air Temp. (°C)
0	16.0	0	18.0	0	18.0
100	15.0				
200	15.0				
300	14.0				
400	13.0				
500	12.0				
600	14.0				
700	13.5				
800	13.0				
900	12.0				
1000	12.0				
1100	11.0				
1200	10.0				

FIGURE 6.2 ELR AND LAPSE RATES FOR TWO RISING PARCELS OF AIR

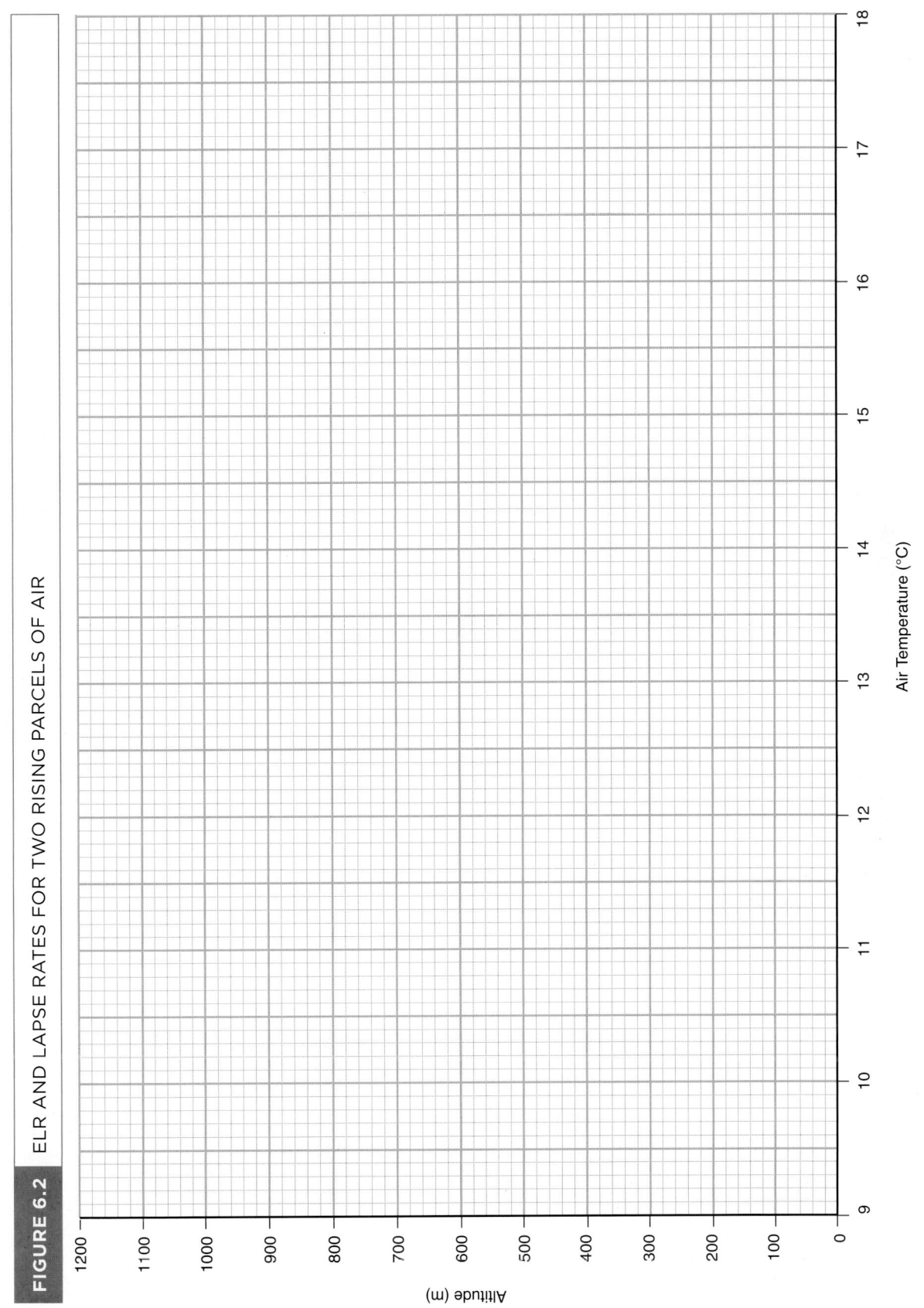

4. a. Is the density of parcel A greater at the earth's surface or at 1200 meters?

   b. Why?

5. a. At 1200 meters, is the air surrounding parcel B more, less, or the same density as the air *inside* parcel B?

   b. Why?

6. a. Which parcel produced the cloud with the lowest (nearest to the surface) cloud base?

   b. Why?

7. a. Which parcel produced the cloud with the most vertical development?

   b. Why?

8. Which parcel is more likely to produce precipitation?

Name: Section:

PART 2 † ADIABATIC PROCESSES AND OROGRAPHIC PRECIPITATION

Figure 6.3 shows a parcel of air starting at point A, which is at sea level (0 meters). The temperature of the air at point A is 25°C and the water vapor content is 15 g/kg. Refer to Figure 6.1 (in Part 1) for information on the saturation mixing ratio. The parcel of air is forced to rise over the mountain, eventually ending at point D, which is also at sea level (0 meters).

FIGURE 6.3 OROGRAPHIC PRECIPITATION

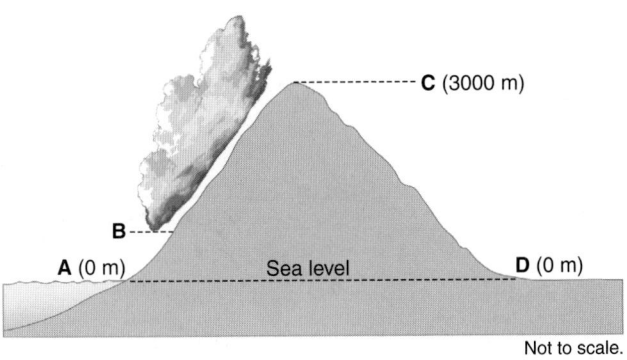

Not to scale.

1. What is the elevation of position B, the base of the cloud?

2. What is the temperature of the parcel of air at the top of the mountain, position C, assuming the height of the mountain is 3000 meters?

3. What is the temperature of the parcel at position D, assuming that all the moisture was lost on the windward side of the mountain and the air stayed below the saturation point as it sank down the leeward side?

4. a. Is the temperature at position D the same, higher than, or lower than the temperature at position A?

   b. Why?

5. a. As the air subsides down the leeward side of the mountain will the relative humidity increase, decrease, or stay the same?

   b. Why?

6. What is the name for the locally generated warm, dry wind that descends the leeward slope of mountains?

Name: _____   Section: _____

PART 3 † INSTABILITY AND CONVECTIONAL RAINFALL

Figure 6.4 shows a typical scenario of convectional uplift in coastal areas. Think of the air over the surface labeled "Ocean" as the air of the surrounding environment (or still air), which is cooling at an environmental lapse rate (ELR) of 0.52°C/100 meters. Over the beach the land is heating the air, causing convectional uplift. The air rises and cools at the adiabatic lapse rate of temperature. Recall that the rate at which the air cools depends on whether the air is saturated or unsaturated. Use the information provided about the air above the ocean (still air) and the air over the beach (rising air due to convection) shown in Figure 6.4 to answer the following questions.

FIGURE 6.4 CONVECTIONAL PRECIPITATION

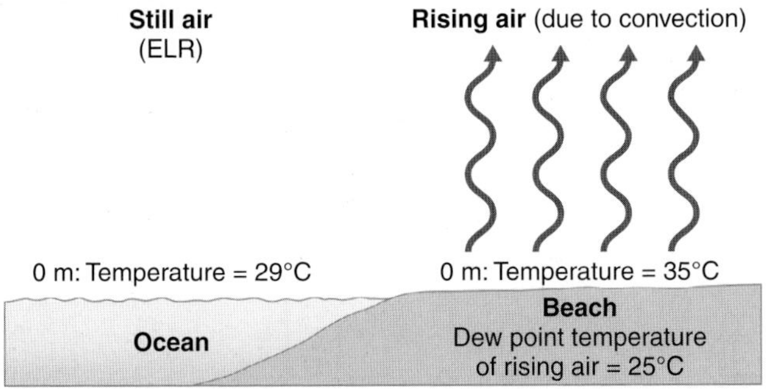

Still air (ELR)

Rising air (due to convection)

0 m: Temperature = 29°C   0 m: Temperature = 35°C

Ocean   Beach
Dew point temperature of rising air = 25°C

1. In this scenario, what is the value of the:

   Environmental lapse rate = _____

   Dry adiabatic lapse rate = _____

   Saturated adiabatic lapse rate = _____

2. a. Draw a graph of the ELR from 0 meters up to an elevation of 3000 meters to represent the still air over the ocean. Use the graph paper provided (Figure 6.5).

   b. Draw a graph of the adiabatic lapse rates of temperature (dry and saturated) from 0 meters up to an elevation of 3000 meters for the rising air over the land. Use the same graph paper as the ELR (Figure 6.5).

3. Label the DALR and SALR portions of the adiabatic lapse rate graph.

4. a. Label the portion of the atmosphere where the rising air is unstable.

   b. Label the portion of the atmosphere where the rising air is stable.

5. At what elevation will the base of the convectional cloud form?

6. What is the elevation of the top of the cloud?

7. What is the air temperature at the top of the cloud?

FIGURE 6.5 ELR AND LAPSE RATE FOR RISING AIR AT THE BEACH

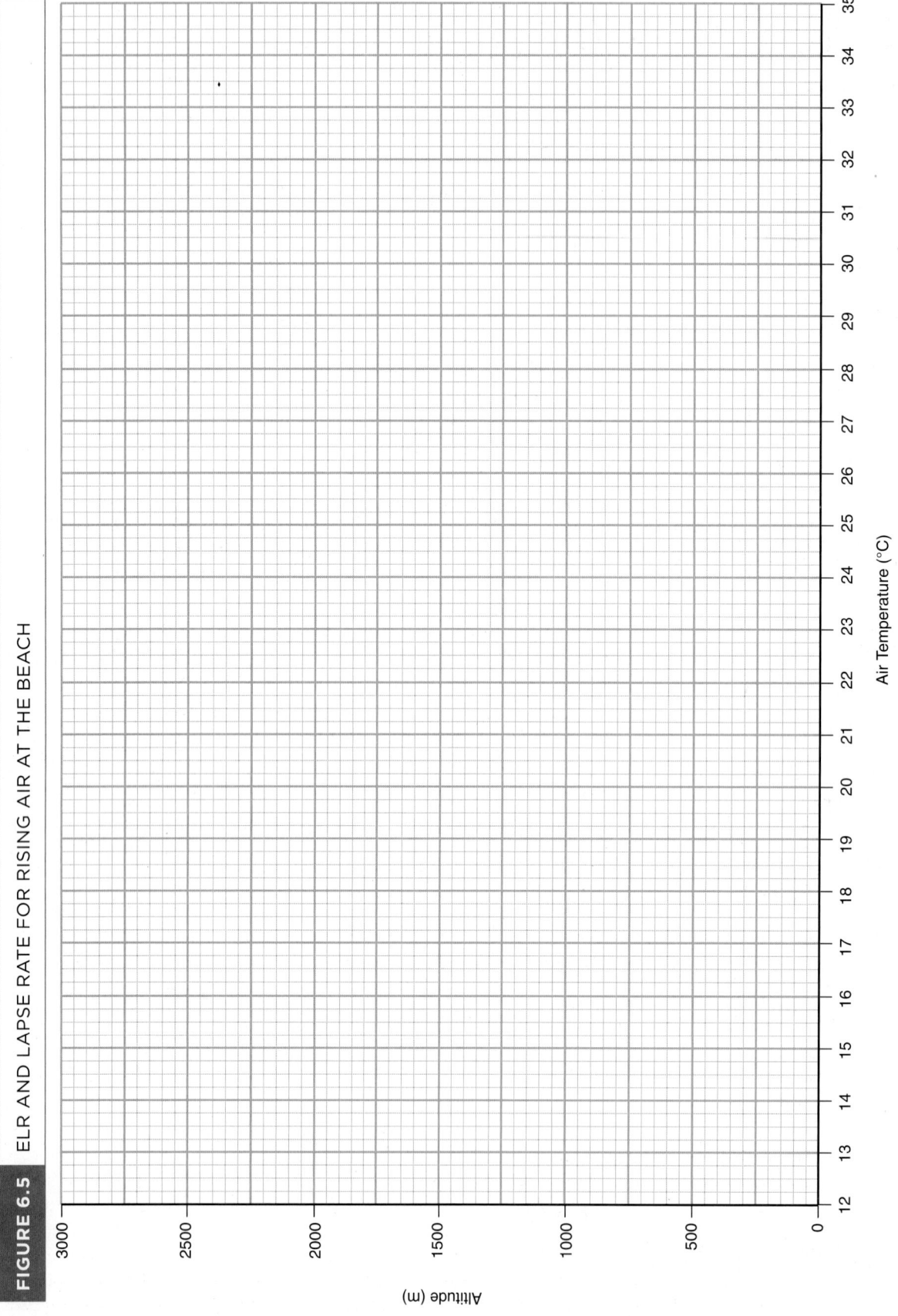

EXERCISE 7 ǀ MIDLATITUDE WEATHER AND WEATHER MAP INTERPRETATION

PURPOSE

The purpose of this exercise is to develop an understanding of the elements that control the weather of the midlatitudes, especially those elements found in North America, and to read and interpret weather maps.

LEARNING OBJECTIVES

By the end of this exercise you should be able to

- interpret atmospheric conditions from a weather map;
- explain how air mass characteristics change with season; and
- predict how weather conditions change as fronts move through.

INTRODUCTION

Our weather is in a constant state of change, being controlled by a complex set of interactions between several atmospheric elements. Of great importance to the weather of the midlatitudes is the influence of migratory high and low pressure cells that form along the polar front. These cyclonic and anticyclonic systems, along with their associated air masses and fronts, result in highly variable weather in the midlatitudes and are a prominent feature of midlatitude climates. The movement of these pressure systems largely determines the day-to-day changes in atmospheric conditions that we call "weather." Hence, an understanding of the internal characteristics of midlatitude cyclones and fronts is important in understanding the weather and climate of the midlatitudes.

Fronts

Fronts are a common feature of a midlatitude cyclone. A **front** is a boundary between contrasting air masses. Because fronts often form where convergence of air masses results in rising air, fronts are located in **troughs** of low pressure. **Cold fronts** form where cold air replaces warm air at a given location. **Warm fronts** form where warm air replaces cold air at a given location. **Stationary fronts** do not exhibit much horizontal movement. **Occluded fronts** form by cold fronts "overrunning" surface warm fronts. Stormy weather is quite common along or ahead of fronts. Violent weather is usually associated with cold fronts because of their steep frontal surface and faster rate of movement compared to warm fronts. As a result, thunderstorms are commonly associated with cold fronts, especially in summer, and the precipitation associated with cold fronts is usually intense but short in duration. Warm front precipitation, in contrast, tends to be longer in duration but generally less intense. Stationary and occluded fronts can have a wide range of weather conditions associated with them.

Precipitation

Recall that the formation of precipitation begins with the uplift of air, which in turn causes cooling and condensation of water vapor. Uplift and associated cooling are accomplished in four ways:

1. **Frontal uplift**—uplift along a weather front where air masses with different temperature and moisture characteristics collide;
2. **Convergent uplift**—uplift due to converging air streams associated with areas of low pressure;
3. **Orographic uplift**—the forced ascent of air along the windward slope of mountains; and
4. **Convectional uplift**—uplift due to heating of air near the surface with a subsequent decrease in air density and increase in air buoyancy.

A number of factors influence the spatial distribution of precipitation. One of the most important is the dominance of air masses at a place. An **air mass** is a vast pool of air that is fairly homogeneous with respect to its temperature and moisture content throughout its horizontal extent. The cold, continental polar (cP) air masses that form over subarctic continents contain little moisture. The warm, moist, maritime tropical (mT) air masses that originate over the subtropical oceans contain much more moisture and serve as the source of atmospheric moisture for precipitation in the midlatitudes. Fronts develop where contrasting air masses meet. The midportion of the United States has been characterized as the "battle ground of air masses" as warm moist air from the Gulf of Mexico collides with the cooler and drier air masses from the Pacific and arctic Canada. The **polar front** forms between tropical-type air and polar-type air and shifts from about 35° latitude in the winter to 55° latitude in the summer.

59

Many of our major midlatitude storm systems are born along the polar front.

Reading and Interpreting Weather Maps

At regular intervals each day, weather observers throughout the world complete a series of meteorological observations and record conditions such as wind direction and velocity, atmospheric pressure and temperature, dew point, sky cover, and cloud types. This information is encoded and relayed to central weather offices. There, the information is plotted on weather maps, either in code or "plain language" actual values, and input into sophisticated computer models. Weather analysts then determine the general weather conditions by interpreting the mapped data for each station and the results from the computer analysis. Figure 7.1 provides an example of a simplified weather station model. Consult your textbook for a more complete example.

Weather station model information is recorded using shorthand notation. In Figure 7.1, the circle identifies the location of a weather station. The present air temperature, in degrees Fahrenheit, is located in the upper left corner (42°F). Dew point temperature, the temperature to which the air must be cooled for saturation, is located in the lower left corner beneath the air temperature (29°F). Wind direction is indicated with a line radiating out from the circle. The wind blows along the line *toward* the circle, thus the wind direction in Figure 7.1 is southeast. Remember, winds are named for where they come *from*. Wind speed is represented with a series of feathers or triangles at the end of the wind direction line. Table 7.1 gives the various symbols used to illustrate wind speed. Note that each symbol represents a range of wind speed values, therefore, the convention that a half feather represents approximately 5 knots, a full feather 10 knots, and a triangle 50 knots is sufficient. The station model presented in Figure 7.1 indicates a wind of about 5 knots.

The present atmospheric pressure (to the tenths of millibars) is given in the upper right corner of the model. With the shorthand notation used, the leading 9 or 10 has been left off the pressure reading, and the decimal point has not been printed. To interpret pressure data from a weather map correctly, use the following convention:

if the first digit is ≥ 5, add a 9 in front or,

if the first digit is < 5, add a 10 in front.

For example, the number 100 would represent an actual pressure reading of 1010.0 millibars. A value of 999 would really be 999.9 millibars. In any case, remember to put a decimal point in front of the last digit. On the station model shown in Figure 7.1, the air pressure is 1021.4 mb. Cloud cover is reported by shading in various portions of the station circle. Table 7.2 summarizes the symbols used for cloud cover. The station model in Figure 7.1 indicates one-tenth or less of the sky is covered.

Weather maps contain a variety of additional weather symbols (see Table 7.2). **Isobars** (lines of equal pressure) show the distribution of surface air pressure. Centers of high and low pressure are depicted by using a letter symbol, *L* for low pressure, *H* for high pressure, or by simply writing the word "Low" or "High." Areas of precipitation are shaded on the weather map. The positions of fronts are indicated with a line on which symbols are attached to differentiate between frontal types. Cold fronts are symbolized with triangles aligned and pointing in the direction the front is moving. Filled semi-circles are used to indicate the position of a warm front. Occluded fronts are shown with alternating warm and cold front symbols aligned on one side of the line and pointing in the direction of frontal movement. Alternating warm and cold front symbols on opposite sides of a line indicate the position of a stationary front. Air mass types may also be indicated on weather maps. The shorthand notation is to use lower case letters to represent moisture conditions of the source area, c for continental or m for maritime, followed by upper case letters to represent the temperature conditions of the source area, P for polar or T for tropical (Table 7.3). Thus, an air mass signified as mP is a maritime polar (cold, moist) air mass.

FIGURE 7.1	SIMPLIFIED WEATHER STATION MODEL

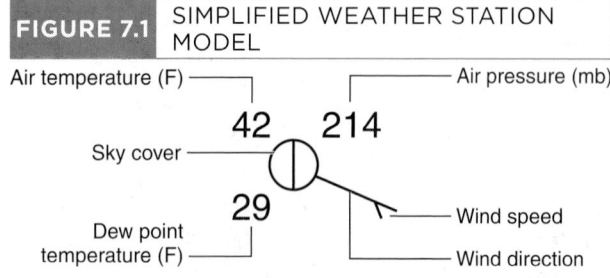

TABLE 7.1 WIND SPEED SYMBOLIZATION

Symbol	Statute miles per hour	Knots	Kilometers per hour
⊚	Calm	Calm	Calm
—	1–2	1–2	1–3
⌐	3–8	3–7	4–13
\	9–14	8–12	14–19
\\	15–20	13–17	20–32
\\\	21–25	18–22	33–40
\\\\	26–31	23–27	41–50
\\\\\	32–37	28–32	51–60
\\\\\\	38–43	33–37	61–69

Symbol	Statute miles per hour	Knots	Kilometers per hour
\\\\\\\	44–49	38–42	70–79
\\\\\\\\	50–54	43–47	80–87
▲	55–60	48–52	88–96
▲\	61–66	53–57	97–106
▲\\	67–71	58–62	107–114
▲\\\	72–77	63–67	115–124
▲\\\\	78–83	68–72	125–134
▲\\\\\	84–89	73–77	135–143
▲▲	119–123	103–107	144–196

TABLE 7.2 SELECTED WEATHER MAP SYMBOLS

Symbol	Description
◠◠◠	Warm front (moving ↑)
▲▲▲	Cold front (moving ↑)
▲◠▲◠	Occluded front (moving ↑)
▼◠▼◠	Stationary front (no movement)
▬	Area of precipitation

Symbol	Sky cover
○	No clouds
⊘	Less than one-tenth to one-tenth
◔	Two-tenths or less than three-tenths
◔	Four-tenths
◐	Five-tenths
◑	Six-tenths
◕	Seven-tenths or eight-tenths
◕	Nine-tenths or overcast with openings
●	Completely overcast
⊗	Sky obscured

TABLE 7.3	AIR MASSES
cA: continental Arctic	mE: maritime Equatorial
cP: continental Polar	mP: maritime Polar
cT: continental Tropical	mT: maritime Tropical

IMPORTANT TERMS, PHRASES, AND CONCEPTS

front
trough
cold front
warm front
stationary front
occluded front
frontal uplift

convergent uplift
orographic uplift
convectional uplift
air mass
polar front
isobars

Name: _____ Section: _____

WEATHER MAP INTERPRETATION

1. Use the July 1, 1991 weather map (Figure 7.3) to obtain the air temperature and air pressure at each of the cities listed in Table 7.4.

2. a. Plot the air temperature for each city in Table 7.4 on the graph in Figure 7.2 and connect the points with a smooth red line.

   b. Plot the air pressure for each city in Table 7.4 on the graph in Figure 7.2 and connect the points with a smooth blue line.

TABLE 7.4 TEMPERATURE AND PRESSURE FOR FIVE CITIES

City (Graph Symbol)	Temperature (°F)	Pressure (mb)
Pocatello, Idaho (ID)		
Rock Springs, Wyoming (WY)		
Denver, Colorado (CO)		
Dodge City, Kansas (KS)		
Lake Charles, Louisiana (LA)		

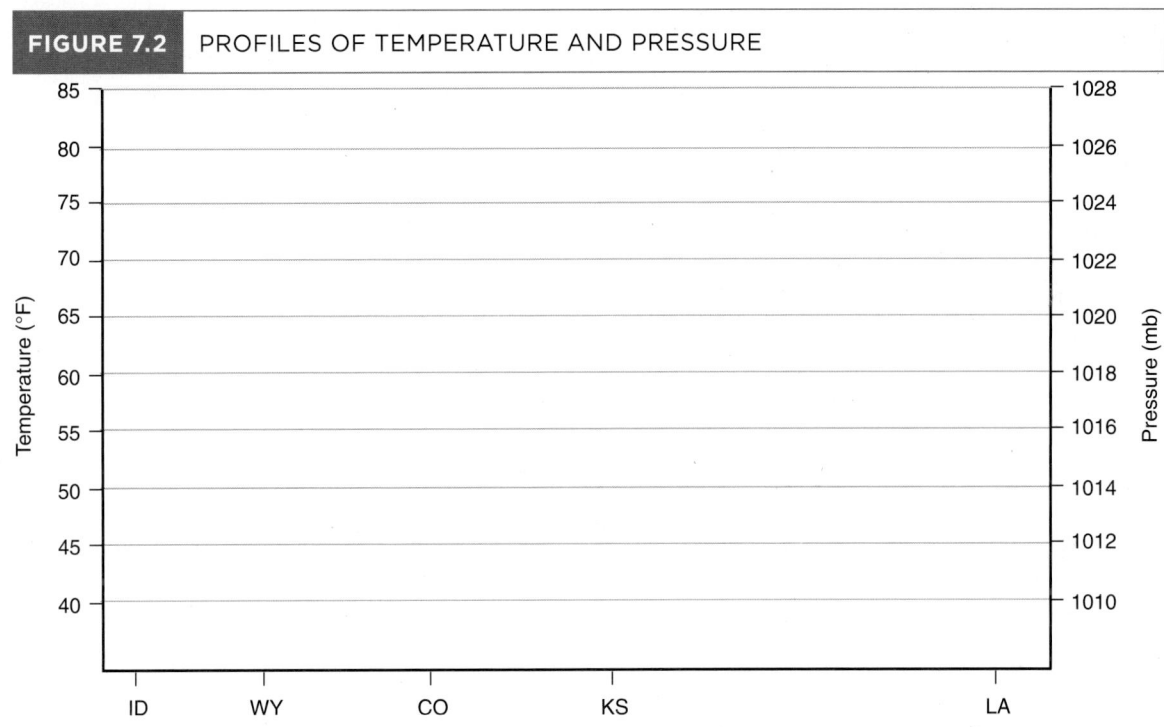

FIGURE 7.2 PROFILES OF TEMPERATURE AND PRESSURE

3. a. What kind of front is found between Pocatello and Lake Charles on the July 1, 1991, weather map (Figure 7.3)?

   b. What direction is this front moving toward?

FIGURE 7.3 JULY 1, 1991, WEATHER MAP

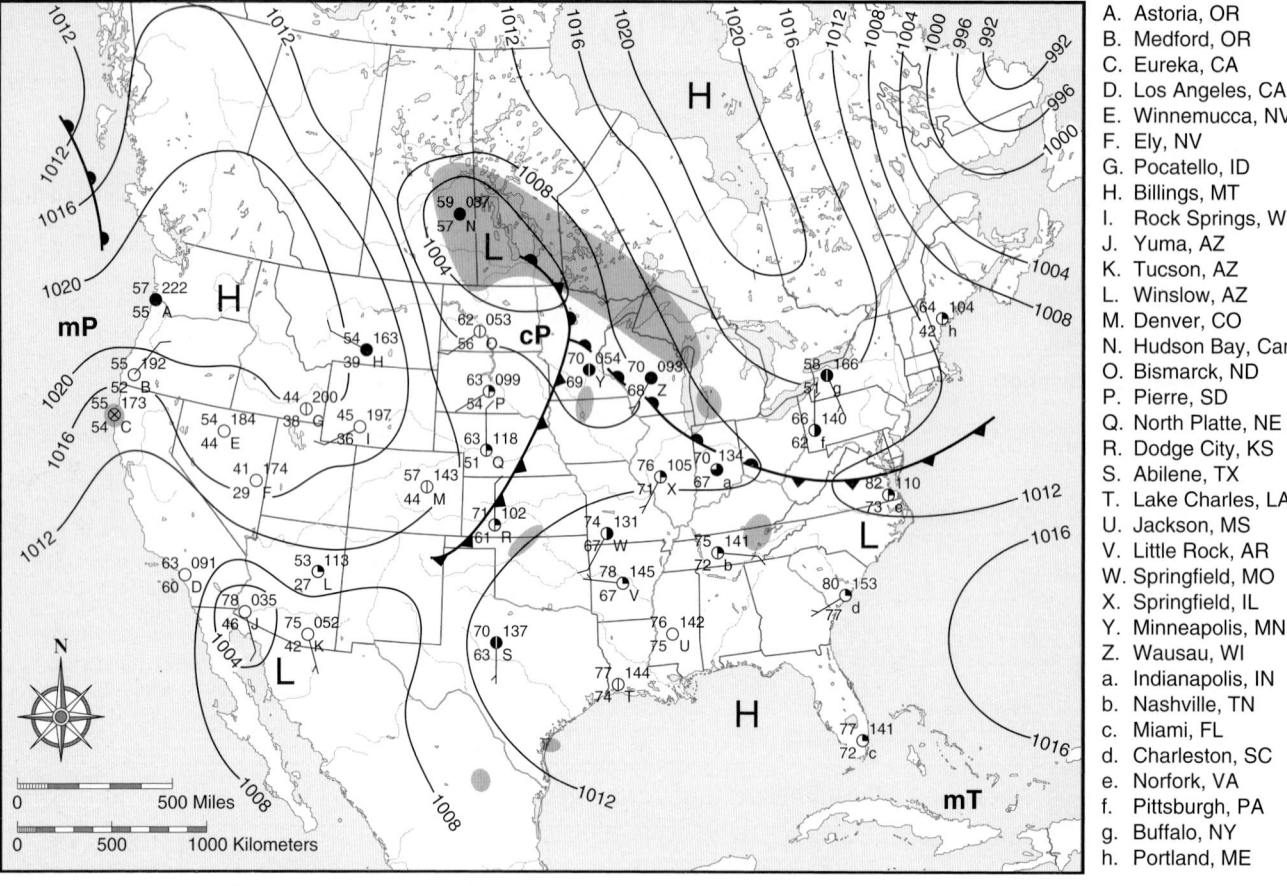

A. Astoria, OR
B. Medford, OR
C. Eureka, CA
D. Los Angeles, CA
E. Winnemucca, NV
F. Ely, NV
G. Pocatello, ID
H. Billings, MT
I. Rock Springs, WY
J. Yuma, AZ
K. Tucson, AZ
L. Winslow, AZ
M. Denver, CO
N. Hudson Bay, Can
O. Bismarck, ND
P. Pierre, SD
Q. North Platte, NE
R. Dodge City, KS
S. Abilene, TX
T. Lake Charles, LA
U. Jackson, MS
V. Little Rock, AR
W. Springfield, MO
X. Springfield, IL
Y. Minneapolis, MN
Z. Wausau, WI
a. Indianapolis, IN
b. Nashville, TN
c. Miami, FL
d. Charleston, SC
e. Norfolk, VA
f. Pittsburgh, PA
g. Buffalo, NY
h. Portland, ME

Source: Data from National Weather Service

    c. Based on the position of this front on the weather map, draw its approximate position on the graph in Figure 7.2 using the correct weather map symbol.

4. a. Examine the graph of air pressure in Figure 7.2. Where is air pressure lowest?

    b. Why?

5. What kind of front stretches from Indianapolis, Indiana, northwestward to Minneapolis, Minnesota, on the July 1, 1991, map (Figure 7.3)?

6. Note the areas of precipitation on the Gulf Coast of Florida, over Lake Superior, and over Hudson Bay, Saskatchewan (letter *N*) on the July 4, 1991, weather map (Figure 7.5). What were the likely uplift mechanisms that caused precipitation (gray shading) at these three locations?

    Gulf Coast of Florida _____

    Lake Superior _____

    Hudson Bay (letter *N*) _____

7. Use the July 3, 1991, and January 3, 1988, maps (Figures 7.4 and 7.6) to fill in Table 7.5.

FIGURE 7.4　JULY 3, 1991, WEATHER MAP

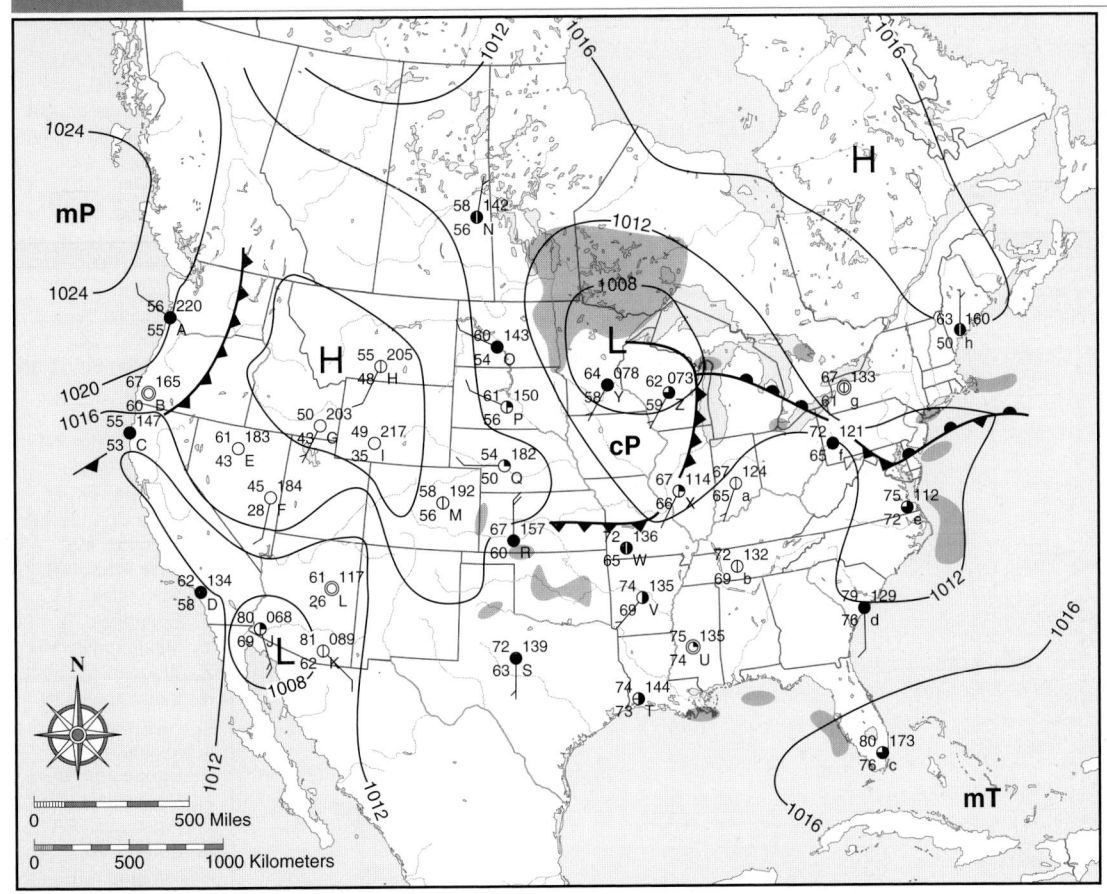

A. Astoria, OR
B. Medford, OR
C. Eureka, CA
D. Los Angeles, CA
E. Winnemucca, NV
F. Ely, NV
G. Pocatello, ID
H. Billings, MT
I. Rock Springs, WY
J. Yuma, AZ
K. Tucson, AZ
L. Winslow, AZ
M. Denver, CO
N. Hudson Bay, Can
O. Bismarck, ND
P. Pierre, SD
Q. North Platte, NE
R. Dodge City, KS
S. Abilene, TX
T. Lake Charles, LA
U. Jackson, MS
V. Little Rock, AR
W. Springfield, MO
X. Springfield, IL
Y. Minneapolis, MN
Z. Wausau, WI
a. Indianapolis, IN
b. Nashville, TN
c. Miami, FL
d. Charleston, SC
e. Norfork, VA
f. Pittsburgh, PA
g. Buffalo, NY
h. Portland, ME

Source: Data from National Weather Service

TABLE 7.5　SUMMER AND WINTER AIR MASS CHARACTERISTICS

Station / Map	Air Temperature	Dew Point	Air Mass
Bismarck, North Dakota (ND) 1/3/88			
Miami, Florida (FL) 1/3/88			
Astoria, Oregon (OR) 1/3/88			
Miami, Florida (FL) 7/3/91			
Astoria, Oregon (OR) 7/3/91			
Bismarck, North Dakota (ND) 7/3/91			

8. Using your knowledge of air masses and the data in Table 7.5 as a guide, compare and contrast the temperature and humidity characteristics of cP, mP, and mT air masses during the summer and winter.

   a. cP air mass

   b. mP air mass

   c. mT air mass

65

FIGURE 7.5 JULY 4, 1991, WEATHER MAP

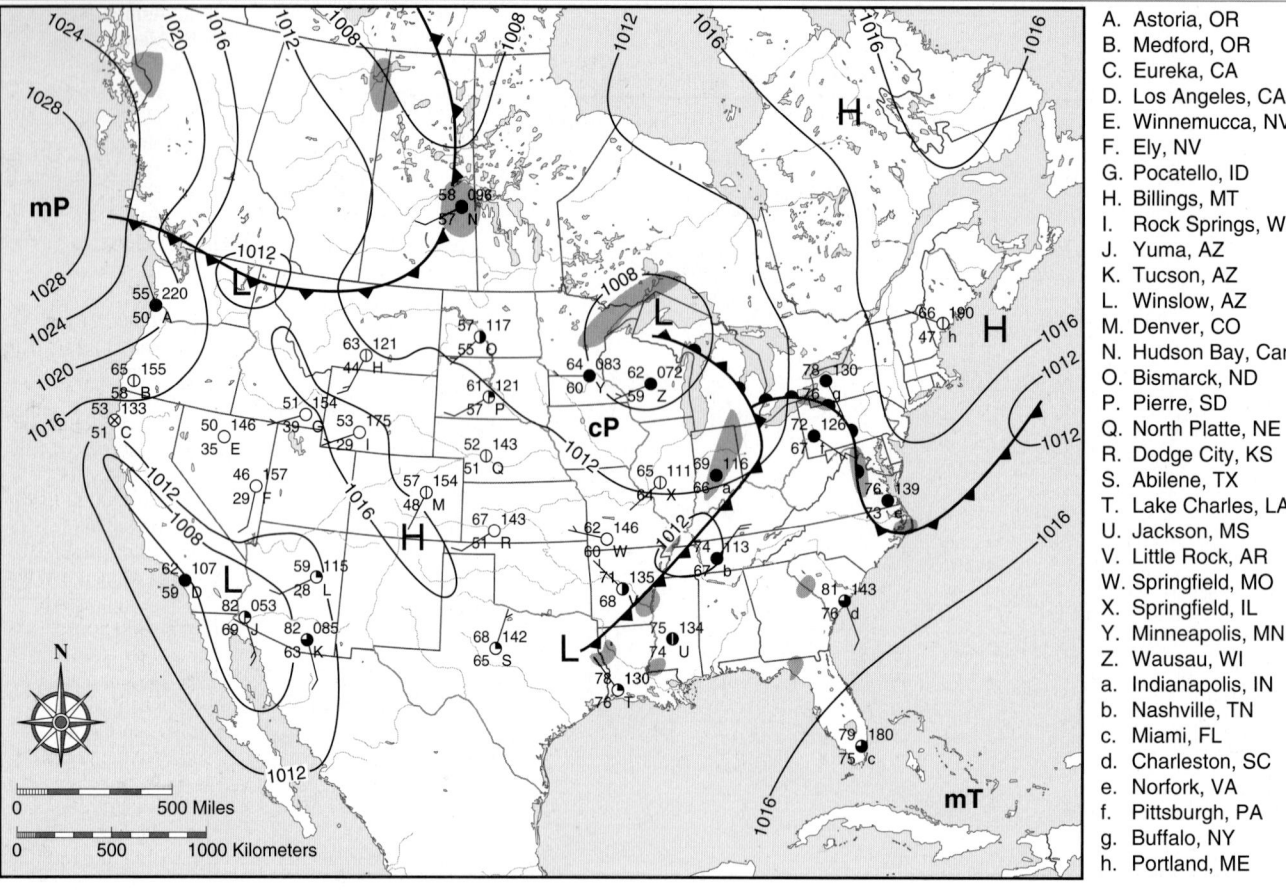

A. Astoria, OR
B. Medford, OR
C. Eureka, CA
D. Los Angeles, CA
E. Winnemucca, NV
F. Ely, NV
G. Pocatello, ID
H. Billings, MT
I. Rock Springs, WY
J. Yuma, AZ
K. Tucson, AZ
L. Winslow, AZ
M. Denver, CO
N. Hudson Bay, Can
O. Bismarck, ND
P. Pierre, SD
Q. North Platte, NE
R. Dodge City, KS
S. Abilene, TX
T. Lake Charles, LA
U. Jackson, MS
V. Little Rock, AR
W. Springfield, MO
X. Springfield, IL
Y. Minneapolis, MN
Z. Wausau, WI
a. Indianapolis, IN
b. Nashville, TN
c. Miami, FL
d. Charleston, SC
e. Norfork, VA
f. Pittsburgh, PA
g. Buffalo, NY
h. Portland, ME

Source: Data from National Weather Service

9. Why are air masses usually warmer and more humid in the summer than in the winter?

10. a. Which air mass has more water vapor in it, mP or mT? _____

    b. How do you know?

11. a. Which city, Miami or Astoria (Table 7.5), had the higher relative humidity in January?

    b. How do you know?

66

FIGURE 7.6 JANUARY 3, 1991, WEATHER MAP

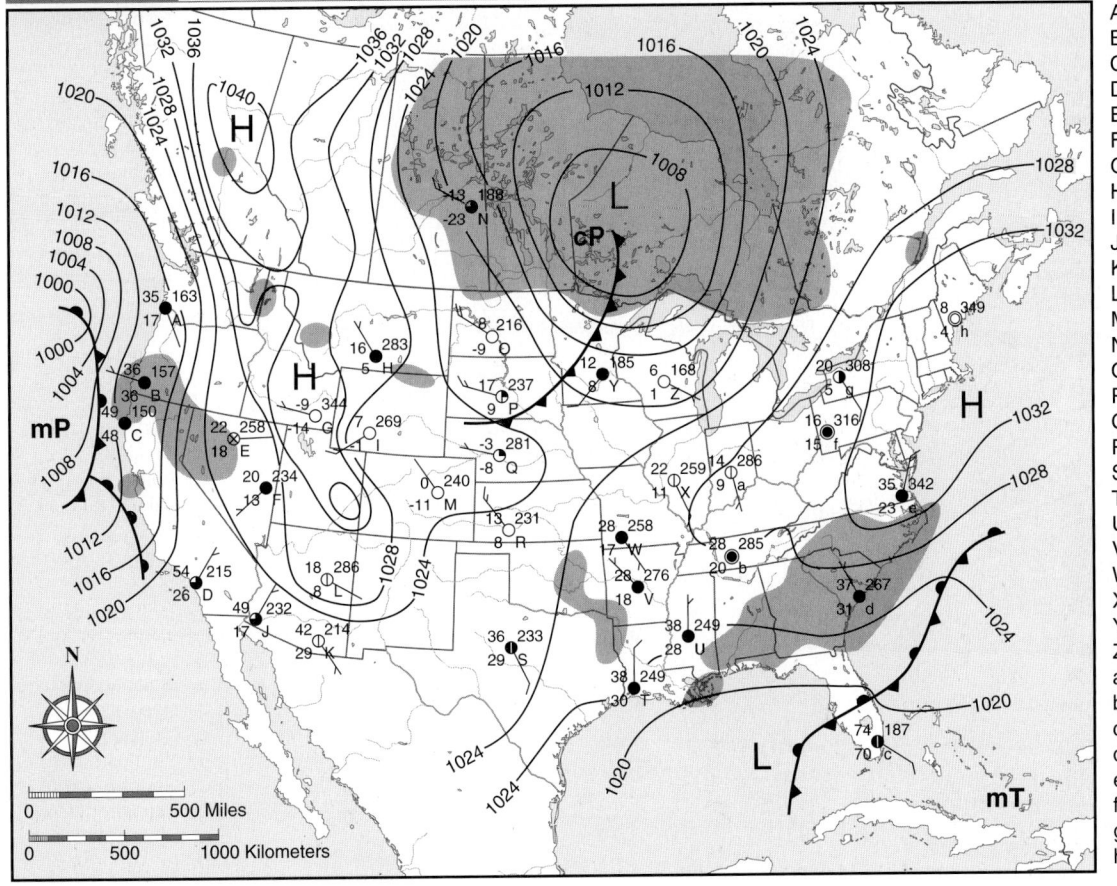

A. Astoria, OR
B. Medford, OR
C. Eureka, CA
D. Los Angeles, CA
E. Winnemucca, NV
F. Ely, NV
G. Pocatello, ID
H. Billings, MT
I. Rock Springs, WY
J. Yuma, AZ
K. Tucson, AZ
L. Winslow, AZ
M. Denver, CO
N. Hudson Bay, Can
O. Bismarck, ND
P. Pierre, SD
Q. North Platte, NE
R. Dodge City, KS
S. Abilene, TX
T. Lake Charles, LA
U. Jackson, MS
V. Little Rock, AR
W. Springfield, MO
X. Springfield, IL
Y. Minneapolis, MN
Z. Wausau, WI
a. Indianapolis, IN
b. Nashville, TN
c. Miami, FL
d. Charleston, SC
e. Norfork, VA
f. Pittsburgh, PA
g. Buffalo, NY
h. Portland, ME

Source: Data from National Weather Service

12. a. Where will it feel more humid, Miami or Astoria (Table 7.5)?

    b. How do you know?

    c. If intensity is related to the pressure gradient across the system, how can you determine the intensity of a cyclone from a weather map?

13. Intense cyclones are characterized by their strong pressure gradients, high winds, and severe weather, be this blizzard conditions during the winter or thunderstorm and tornado conditions during the spring and summer.

    a. Based on the weather maps in Figures 7.3 to 7.7, during which season (summer or winter) are the surface cyclones most intensely developed?

    b. Why is this so?

FIGURE 7.7   JANUARY 4, 1991, WEATHER MAP

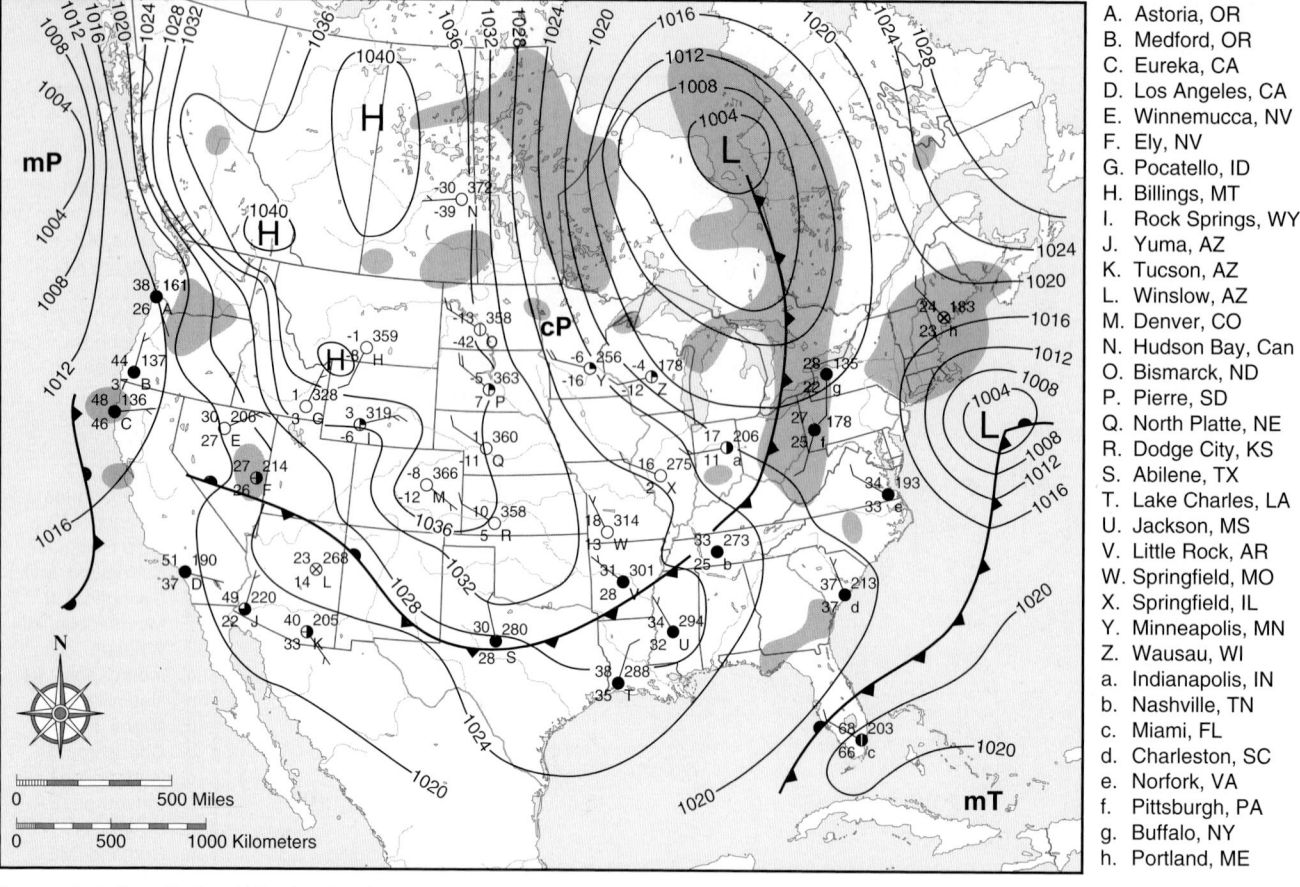

Source: Data from National Weather Service

14. On January 3, 1988 (Figure 7.6), a cold front stretching from central Canada to Pierre, South Dakota, is moving toward the _____.

15. Bismarck, North Dakota, lies to the west of this cold front and Minneapolis, Minnesota, to the southeast. Collect the following weather data from the January 3, 1988, map (Figure 7.6) for these two cities located on opposite sides of the front.

16. In 24 hours the front is expected to pass Minneapolis, Minnesota. Examine the conditions on both sides of the front as represented in the data compiled in Table 7.6. Predict what will happen to the air tempaerature (increase or decrease), dew point temperature (increase or decrease), sky cover (increase or decrease), and change in wind direction as the front moves through Minneapolis.

TABLE 7.6 WEATHER CONDITIONS ON OPPOSITE SIDES OF A COLD FRONT

Weather Element	Bismarck	Minneapolis
Air temperature		
Dew point temperature		
Wind direction		
Wind speed		
Sky cover		
Air pressure		

17. Collect the following data from the January 4, 1988, map (Figure 7.7).

TABLE 7.7 WEATHER CONDITIONS AT MINNEAPOLIS

Weather Element	Minneapolis	Trend over the past 24 hours (e.g., increase or decrease)
Air temperature		
Dew point temperature		
Wind direction		
Wind speed		
Sky cover		
Air pressure		

18. Compare the actual conditions to your prediction.

EXERCISE 9 † SOIL MOISTURE BUDGETS

PURPOSE

The purpose of this exercise is to analyze moisture transfer and the seasonal progression of local moisture balances, and to learn how to interpret summary values and graphs of soil moisture variables.

LEARNING OBJECTIVES

By the end of this exercise you should be able to

- account for moisture transfer between the atmosphere and the soil via precipitation, evapotranspiration, soil moisture recharge, and utilization;
- explain the seasonal progression of local moisture balances, such as the occurrence of surpluses and deficits;
- interpret soil moisture summary values and graphs for different environments;
- explain the impact of soil texture on soil moisture budgets; and
- explain the impact of latitude, altitude, and continentality on moisture balances.

INTRODUCTION

Local climates directly affect vegetation patterns because various plants have specific temperature and moisture requirements. The availability of water is a critical factor affecting the distribution and form of organisms. Moisture and energy continuously transfer between the lithosphere, atmosphere, and biosphere in the hydrologic cycle, and the soil interface is a prime transfer site. Thus soil, in addition to being the source of nutrients, also exerts a powerful climatic and hydrologic influence on plant growth.

Whether or not there is sufficient soil moisture for plant growth depends on the amount and distribution of water supplies, the storage capacity of the soil, the proportion of capillary water remaining in the soil, and rates of **evapotranspiration** (the combination of evaporation and transpiration). Air and soil temperatures further restrict moisture availability because most plants must have access to **effective moisture**—accessible fresh water in a liquid state. Snow is not effective moisture because it is not accessible to plants until it melts; thus the availability of this water is deferred until temperatures increase sufficiently.

To simplify this exercise, we will somewhat unrealistically assume that precipitation is the only source of moisture at a site, that there is no deferral of frozen precipitation's availability, and that evapotranspiration can remove all moisture from the soil if necessary. Furthermore, we shall assume that water is allocated to only three recipients, in order of priority:

1. the atmosphere, via evapotranspiration,
2. the soil, via changes of soil moisture storage, and
3. rivers and streams, via surface runoff.

The Soil Moisture Budget Concept

Did you ever wonder how the operators know when to turn on the water in those giant center-pivot irrigation systems? Or when to open the floodgates on dams? Or when to issue forest fire alerts and implement burning restrictions? All of these, and many other environmental decisions, rely on the day-to-day accounting of the water budget technique. Water budgets also provide a means for comparing the seasonal availability of water in different regions as it cycles through the soil.

Water budget tables are ledgers that account for inputs and outputs of water to the soil moisture zone. Comparing water budgets for different locations illustrates the spatial variability in water availability caused by the interaction of water, heat, and soil. Thus, calculation of a water budget requires information on soil texture, temperature, and precipitation at a site.

Soil texture, the proportional mix of particle sizes, affects the amount of water the soil can hold. **Field capacity** is the maximum amount of water the soil can hold, and this storage ceiling affects the local availability of water. In general, coarse-textured soils have low field capacities while fine-textured soils have high field capacities (Table 9.1).

Temperature directly controls the amount of evapotranspiration at a site. **Potential evapotranspiration** (PE) is the maximum amount of water the site would yield to the atmosphere given the temperature if there were an unlimited supply of soil moisture for **evaporation** and **transpiration** (release of water vapor by plants). Potential evapotranspiration depends solely upon the amount of heat energy available (indexed by temperature) for evaporation

TABLE 9.1	TYPICAL FIELD CAPACITIES (IN WATER EQUIVALENCY UNITS)			
Coarse Sand	Fine Sand	Sandy Loam	Silty Loam	Clay
(coarsest texture --→ finest texture)				
25 mm	102 mm	152 mm	203 mm	305 mm

and plant growth. Since unlimited water usually is not available, it also becomes necessary to calculate **actual evapotranspiration** (AE), which depends on both the heat energy *and* the amount of water available at a site.

Precipitation is the primary input of water to a site. Precipitation in combination with water stored in the soil determines how much water is available for evapotranspiration, soil moisture storage, and surface runoff.

Moisture-abundant environments are environments where precipitation is greater than potential evapotranspiration (P > PE); thus, actual evapotranspiration equals potential evapotranspiration (AE = PE). These sites may experience surpluses of water, water that cannot be stored in the soil because the soil is already at field capacity, and that cannot be evapotranspired because of insufficient additional heat energy. Surplus water becomes surface runoff and supplies water to rivers and streams.

Moisture-limited environments are environments where precipitation is less than potential evapotranspiration (P < PE). In these environments, actual evapotranspiration may still equal potential evapotranspiration provided that some water is stored in the soil. Once soil moisture storage is depleted, however, actual evapotranspiration is less than potential evapotranspiration (AE < PE), and in this case, a deficit of water occurs. Deficits of water result in **droughts**, and droughts often cause stress in plants.

Some locations, such as rainforests, experience chronic surpluses, while others, such as desert locations, experience chronic deficits. Other locations may experience seasonal surpluses and/or deficits, resulting in distinct wet and dry seasons. Still other regions may experience random temporary surpluses or deficits. The chronic, seasonal, or random occurrence of surpluses and deficits may impact the local ecosystem, depending on how well plants are adapted to withstand either surpluses or droughts.

The Soil Moisture Budget Variables

Table 9.2 is an example of a blank water budget table, with columns for each month of the year and for necessary annual totals, and rows for the variables used to calculate monthly moisture balances. The soil moisture variables include

Precipitation (P): the mean monthly total precipitation. Think of P as being the gross moisture income, just as a salary is gross financial income.

Potential Evapotranspiration (PE): the atmospheric moisture demand. C. W. Thornthwaite devised a formula to calculate PE strictly based on temperature (Mather 1974). The higher the temperature, the higher the PE. PE is moisture demand, just as a bill is a monetary demand.

Precipitation-Potential Evapotranspiration (P-PE): the difference between P and PE. If P-PE is negative, that month has insufficient precipitation given the available energy to meet the atmospheric demand. When such a shortfall occurs, the soil must yield moisture (if it is

TABLE 9.2	A BLANK SOIL MOISTURE BUDGET TABLE												
	JAN	FEB	MAR	APR	MAY	JUN	JUL	AUG	SEP	OCT	NOV	DEC	YR
P													
PE													
P-PE													
DST													
ST													
AE													
D													
S													

storing any) to meet the temperature-induced demand. When P-PE is positive, the site automatically fulfills the atmosphere's moisture demand, and the positive value represents leftover moisture that may be allocated to the soil or to surplus (rivers). When allocating this leftover moisture, recharging the soil until it reaches field capacity always takes priority over surplus. P-PE is monthly net moisture, like what remains of a salary after tax withholding and bill payments.

Change in Soil Moisture Storage (ΔST): the amount of water added to or removed from the soil. (The Greek letter delta, Δ, means "change in.") If P-PE is positive, the excess water is added to **recharge** the soil (a positive ΔST in row 4), up to its field capacity. If P-PE is negative, as much soil moisture as necessary to meet PE must be withdrawn from the soil. The withdrawal of water from the soil is called **utilization**. Sometimes all remaining moisture is removed and the soil becomes totally depleted. ΔST is like a transaction note to a bank account, except that there can be only one water budget transaction in any single month. ΔST and P-PE are the only water budget variables that can be negative; be sure to include the signs for them.

Soil Moisture Storage (ST): the amount of water remaining in the soil at the end of each month after each monthly transaction occurs. Soil acts like a savings account for water, but only up to the soil's field capacity, and ST is a current month-end balance for a soil moisture account. Remember that the maximum possible amount of water that the soil can store is its field capacity; *ST cannot exceed field capacity*! ST acts as a running balance from one month to the next; it is the only water budget variable that does so.

$$ST_{current\ month} = ST_{previous\ month} + DST_{current\ month} \quad (9.1)$$

Actual Evapotranspiration (AE): the total amount of moisture (from precipitation and/or soil moisture) that actually evapotranspires during a month. During any wet month (when P > PE), AE automatically equals PE. During any dry month (when P < PE), AE equals precipitation plus the *absolute value* of change in soil moisture storage (P + |ΔST|). Or:

$$\text{if } P \geq PE, \text{ then } AE = PE, \text{ and} \quad (9.2)$$
$$\text{if } P < PE, \text{ then } AE = P + |DST|. \quad (9.3)$$

AE can never exceed PE. AE is like an actual payment made toward bills, even if the AE payment cannot meet the PE demand in full.

Deficit (D): when AE < PE. This occurs when storage (ST) becomes equal to zero. Deficit is always calculated as:

$$D = PE - AE. \quad (9.4)$$

Deficits cannot be "repaid" later once wetter conditions resume, so think of them as defaulted debts.

Surplus (S): when excess water from a positive monthly P-PE still remains unused even after fully recharging soil moisture to field capacity. In other words, for a surplus to occur the following conditions must be met: AE = PE, and ST = field capacity, and some moisture is still unallocated. Surplus is like discretionary income, except that in the real environment any surplus moisture has to go somewhere, so it is permanently lost as runoff to rivers and streams. Deficit and surplus cannot occur simultaneously in the same month!

Working a Soil Moisture Budget: An Example

Vancouver, near the mouth of the Columbia River in Washington State, is a good example for water budget calculations because that site has very pronounced wet and dry seasons. Table 9.3 is a complete water budget table for

TABLE 9.3 SOIL MOISTURE BUDGET FOR VANCOUVER, WASHINGTON

46°N,123°W		Elevation 3 m				Fine sand			Field capacity = 102 mm				
	JAN	FEB	MAR	APR	MAY	JUN	JUL	AUG	SEP	OCT	NOV	DEC	YR
P	146	121	102	69	56	47	31	37	60	116	155	172	1112
PE	11	19	28	41	59	73	85	84	69	47	27	16	559
P-PE	+135	+102	+74	+28	-3	-26	-54	-47	-9	+69	+128	+156	
ΔST	0	0	0	0	-3	-26	-54	-19	0	+69	+33	0	
ST	102	102	102	102	99	73	19	0	0	69	102	102	
AE	11	19	28	41	59	73	85	56	60	47	27	16	522
D	0	0	0	0	0	0	0	28	9	0	0	0	37
S	135	102	74	28	0	0	0	0	0	0	95	156	590

Vancouver. The sequence of numbered steps below "walks through" the entire set of calculations, illustrating rules and pointing out potential pitfalls.

Step 1 The first step is to specify the field capacity according to the soil texture; here it is 102 mm for fine sand (see Table 9.1). Remember later that ST cannot exceed this field capacity!

Step 2 Figure out in what month to start your calculations. Each water budget in this exercise has one storage value with a gray background. Start all your calculations in the following month. For this example, start calculations in October.

Step 3 Calculate P-PE for October. Be sure to include a positive or negative sign. Whenever P-PE is negative, the month is dry; and if P-PE is positive, the month is wet. In Vancouver, for example, March is wet and May is dry.

Step 4 Calculate ΔST for October. Since October's P-PE is positive, and since September's ST < field capacity, all the leftover moisture gets added to the soil; ΔST = +69.

Step 5 Calculate ST for October. Apply the ΔST (change of soil moisture) to the previous month's ST (storage balance) to get the current month's new ST balance (see equation 9.1). For Vancouver, apply the ΔST from October (+69) to September's ST (0), which makes the October ST = 0 + 69 = 69.

Step 6 Calculate AE for October. If P > PE, then AE = PE (equation 9.2). Since October P-PE is positive, AE = PE, thus AE = 47.

Step 7 Calculate D for October. D = PE – AE (equation 9.4). Since P actually yielded enough water to meet the demand for evapotranspiration (AE = PE), Vancouver has no deficit in October (D = 0).

Step 8 Calculate S for October. Since Vancouver's October soil moisture storage (ST = 69) is below field capacity (102 mm), there is no surplus of water (S = 0). This completes October's calculations.

Step 9 Work through November next. One difference from October occurs when calculating ΔST for November. As in October, November's P-PE is positive. Since October's ST < field capacity, some of November's leftover moisture must be added to the soil; however, since ST cannot exceed field capacity, not all of the leftover moisture can be added to the soil. Since October ST = 69, November's ΔST = 33 because 69 + 33 = 102, which is field capacity. Thus ST = 102 in November. The remainder of the positive P-PE will become November's surplus (95).

Step 10 Continue working through each month in a like manner. Since the monthly P and PE values in most water budgets represent climatic normals, refer back to December's ST when calculating January's column. In Vancouver, December's ST of 102 (soil moisture is full) determines January's ΔST of 0 (the full soil could take no more water). Calculations for the months of December through April are all very similar.

Step 11 The month of May is slightly different from the preceding months because May is a dry month (P-PE is negative). When P-PE is negative the rest of the PE demand is met, if possible, by removing water from soil moisture storage (causing a negative ΔST). Since P-PE = –3 in May, ΔST = –3, which means 3 mm of water are removed from the soil (utilized). Since ST = 102 in April, and since ΔST = –3 in May, ST = 102 – 3 = 99 for May (see equation 9.1), thus storage is now less than field capacity. When P-PE is negative, AE = P + |ΔST| (equation 9.3), thus AE = 56 + 3 = 59 in May. Since AE = PE in May, there is no deficit. Since ST is less than field capacity, there is also no surplus in May. In Vancouver soil moisture depletion occurs in May, June, and July, so calculations for June and July are similar to May.

Step 12 The month of August differs from the preceding months because of limited soil moisture storage. As in the prior three months, P-PE is negative and some moisture is stored in the soil. August P-PE = -47. July ST = 19. In May, June, and July, the P-PE difference equaled ΔST, but this can't occur in August because insufficient moisture is available. August ΔST = –19 because that's all the moisture available for removal. This drops ST to zero (ST = 0). AE = P + |ΔST| = 37 + 19 = 56. For the first time, there is a deficit: D = PE-AE = 84 – 56 = 28 (equation 9.4).

Step 13 In September once again, the value for P-PE is negative. When P-PE was negative in August, the last of the water was removed from the soil to make up some (but not all) of the demand, and ST for August became 0. As a result, no further soil water is available to make up September's P-PE difference; so September's ΔST = 0. There was no soil moisture left at the end of August and September added none, so September's ST remains 0. Since no soil water is available in September, AE is only the 60 mm from P, and September's D = 9.

Step 14 Once you have calculated the values for each month, sum and record the annual totals for P, PE, AE, S, and D in the final column. If annual P exceeds annual PE, you have a **wet climate;** if annual PE exceeds annual P, you have a **dry climate.** Vancouver has a wet climate. You do not need annual totals for

P-PE, ΔST, or ST. If the account balances for the entire table then:

$$\text{Annual PE} = \text{Annual AE} + \text{Annual D, and} \quad (9.5)$$

$$\text{Annual P} = \text{Annual AE} + \text{Annual S.} \quad (9.6)$$

Soil Moisture Budget Graphs

Calculating the water budget for a site produces a big table of numbers. Although usable, it is difficult to envision such tabular information rapidly. Graphing table information makes it easier to assimilate and compare information from one or more sites. The graph's vertical scale (Y-axis) indicates water equivalency units. The horizontal scale (X-axis) indicates the months of the year, evenly spaced along the axis.

Figure 9.1's three lines represent monthly potential evapotranspiration (PE), monthly precipitation (P), and monthly actual evapotranspiration (AE). The shadings between these three lines indicate the four **soil moisture seasons**. A positive ΔST indicates soil moisture **recharge;** a negative ΔST shows soil moisture **utilization;** a negative P-PE, ST = 0, *and* AE < PE indicates deficit; and positive P-PE *and* ST = field capacity shows surplus.

Ordinarily, water budget graphs show the soil moisture season shadings, cut proportionally within the month of moisture season transition. The objective of these graphs is to communicate general information about the soil moisture seasons, rather than convey specific point values. Key interpretation criteria are that (1) the *intensity* of a season is indicated by the vertical extent of the shaded area on the graph, and (2) the *duration* of a season is indicated by the horizontal extent of the shaded area on the graph. For Vancouver the most intense soil moisture season is surplus in December, while the shortest season is deficit from mid-August to the end of September.

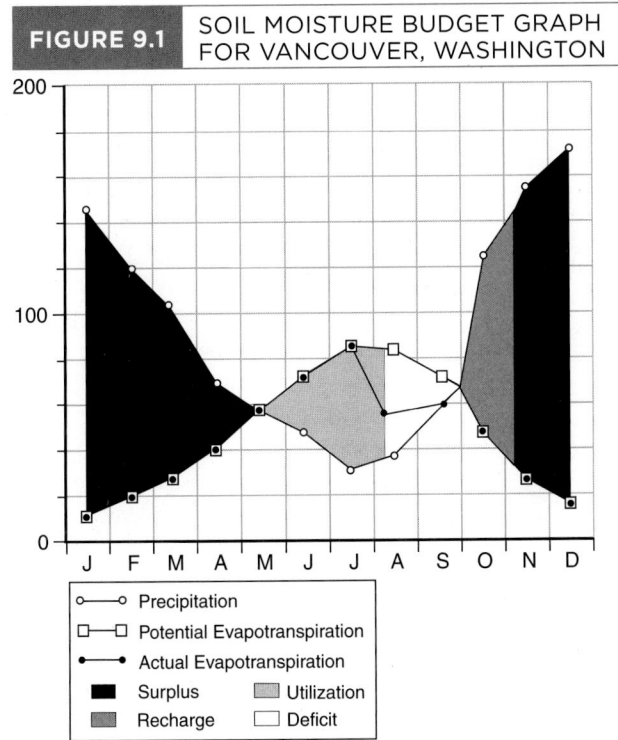

FIGURE 9.1 SOIL MOISTURE BUDGET GRAPH FOR VANCOUVER, WASHINGTON

IMPORTANT TERMS, PHRASES, AND CONCEPTS

soil moisture budget	precipitation (P)
evapotranspiration	recharge
effective moisture	utilization
soil texture	soil moisture storage (ST)
field capacity	deficit (D)
potential evapotranspiration (PE)	surplus (S)
evaporation	wet climate
transpiration	dry climate
actual evapotranspiration (AE)	soil moisture seasons
drought	

REFERENCE

Mather, John. 1974. *Climatology: Fundamentals and Applications*. New York: McGraw-Hill.

Name: _____   Section: _____

PART 1 · SOIL MOISTURE BUDGET TABLES AND GRAPHS

1. Complete the water budget below for Stevens Point, Wisconsin (Table 9.4), a location with a rather different climate from that of Vancouver. Start calculations in the month following the gray background ST value. Whenever necessary, refer to the preceding section explaining the calculations for the Vancouver example. Remember to include the signs in the P-PE and ΔST blanks!

TABLE 9.4 SOIL MOISTURE BUDGET FOR STEVENS POINT, WISCONSIN

45°N, 90°W	Elevation 329 m				Fine sand			Field capacity = _____ mm				
	JAN	FEB	MAR	APR	MAY	JUN	JUL	AUG	SEP	OCT	NOV	DEC	YR
P	28	27	47	70	99	99	90	100	102	60	57	34	813
PE	0	0	0	31	65	91	105	98	72	41	5	0	508
P-PE	+28	+27	+47	+39	+34	+8		+2	+30	+19	+52	+34	
ΔST		0	0	0	0	0		+2	+13	0	0		
ST	102	102	102	102	102	102	87		102	102	102		
AE	0	0	0	31	65	91	105	98	72	41	5	0	508
D	0	0	0	0	0	0	0	0	0	0	0	0	
S	28	27	47	39		8	0	0	17	19	52		

2. Does Stevens Point have a wet or dry climate? How do you know?

3. September's P-PE is +30 mm, but only 13 mm of this returns to the soil as ΔST.

    a. How much of the 30 mm do we still need to account for? _____

    b. Where does this excess moisture go in September?

4. a. What soil moisture season occurs for three quarters of the year in Stevens Point?

    b. What soil moisture season occurs in July at Stevens Point?

101

5. In Figure 9.2, which site has the shortest duration (horizontal extent of shading) of runoff during surplus?

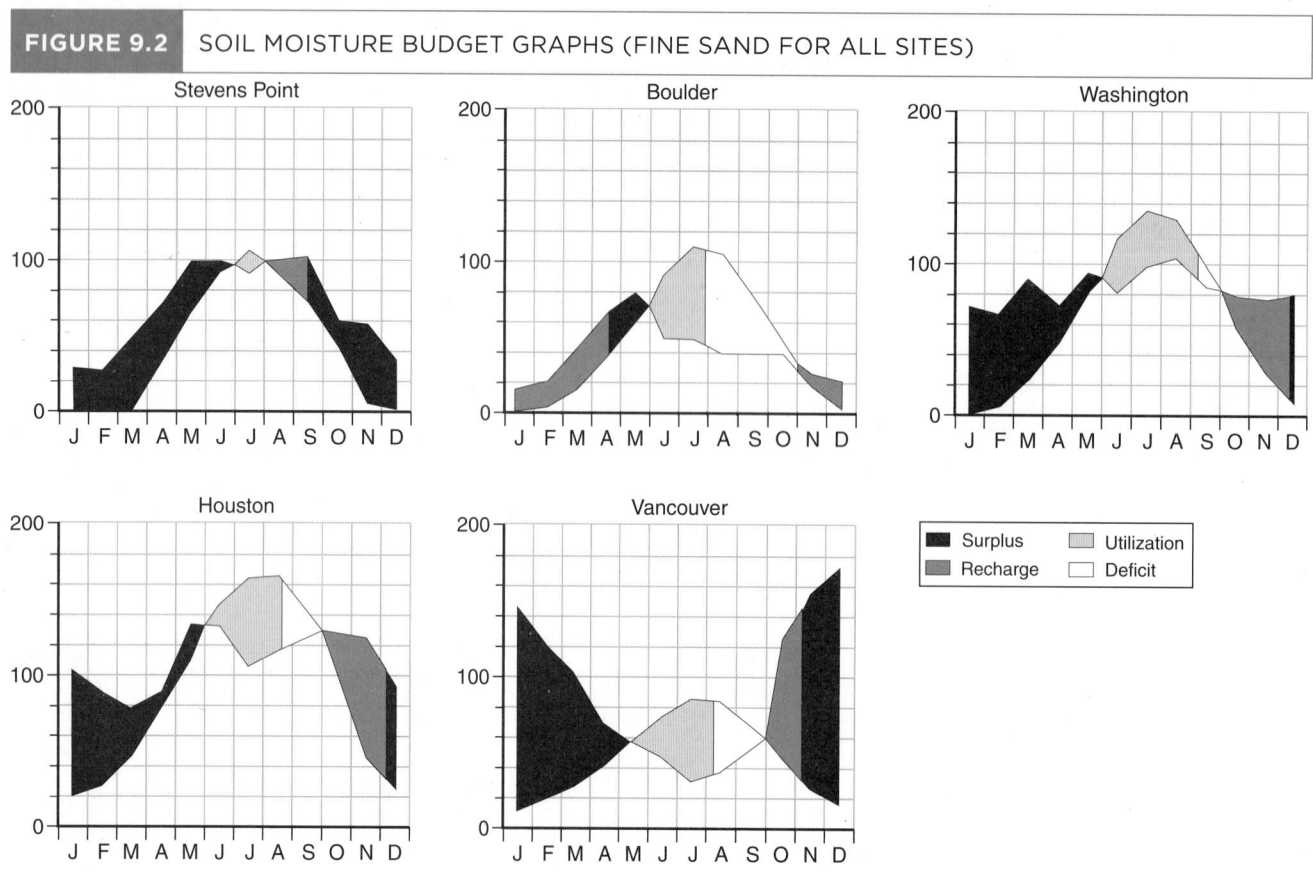

6. In Figure 9.2, which site has the highest intensity (vertical extent of shading) of runoff during surplus?

7. Seemingly at odds with what its graph indicates, there is a creek running through Boulder that has flooded several times in the months of July and August. How can this possibly occur, since Boulder receives insufficient precipitation in July and August to have runoff? Where could the floodwater come from? (Hint: Consult a map of Colorado topography.)

8. Suppose Stevens Point experienced an unusual drought that started in July and lasted through the following February, and that during this time, Stevens Point received no precipitation whatsoever. If normal precipitation resumed in March, why would a drought not necessarily occur the following summer?

9. a. Viewing the maps in Figure 9.3, what states or provinces have the largest differences between annual PE and annual AE?

FIGURE 9.3 ANNUAL SOIL MOISTURE BUDGET VARIABLES MAPS FOR NORTH AMERICA (MM/YR)

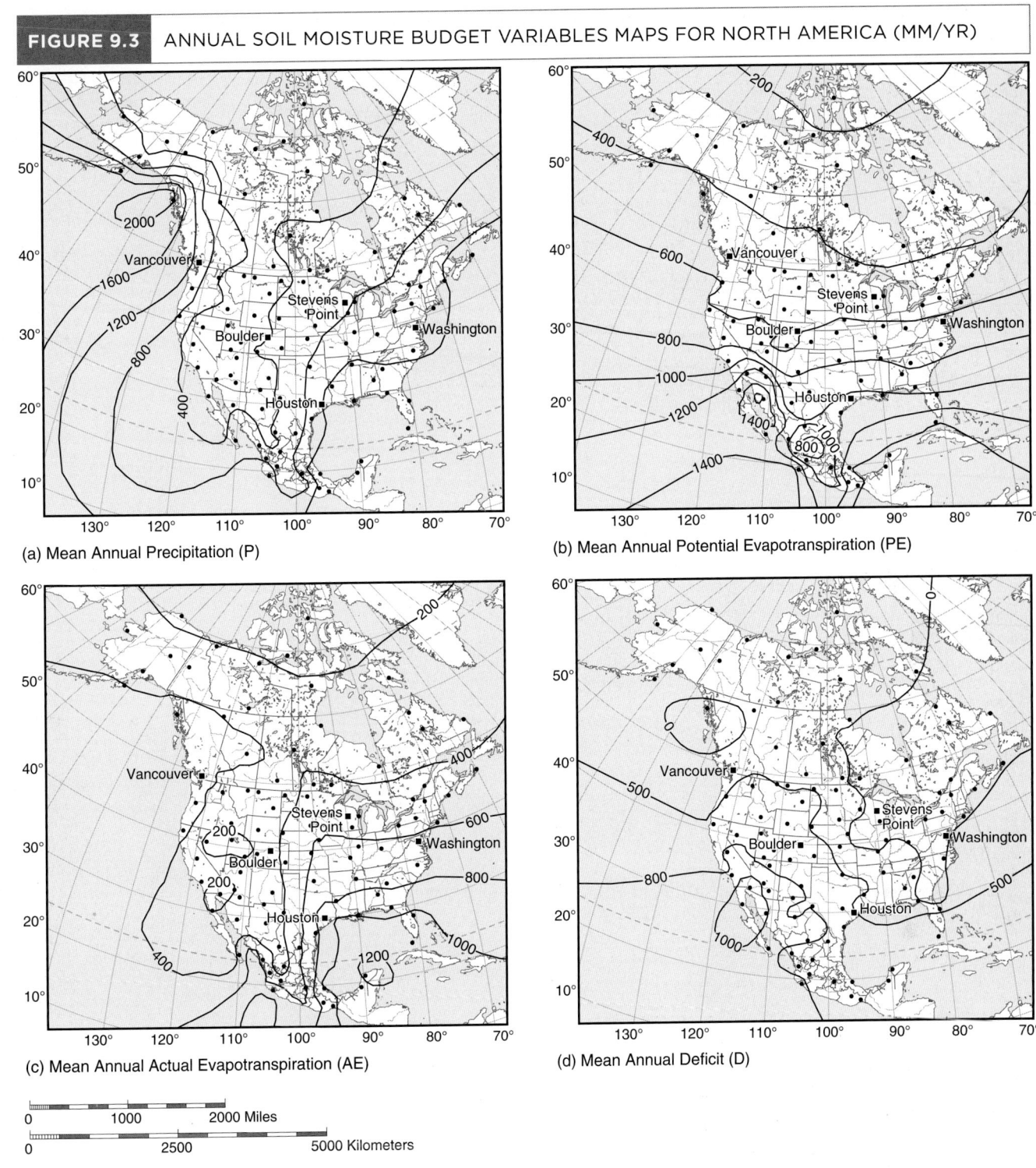

(a) Mean Annual Precipitation (P)
(b) Mean Annual Potential Evapotranspiration (PE)
(c) Mean Annual Actual Evapotranspiration (AE)
(d) Mean Annual Deficit (D)

b. What do the Annual P and Annual D maps in Figure 9.3 indicate as occurring in these areas?

103

c. Which of the five cities shown in Figure 9.3 has the largest difference between annual PE and AE? Refer to their tables (Tables 9.3, 9.4, 9.5a, 9.6, 9.7).

d. Using the graphs in Figure 9.2, see if the city you listed as an answer to part (c) also has the precipitation and deficit conditions that best match your answer to part (b). Yes or no?

Name: _____   Section: _____

PART 2 • COMPARATIVE FIELD CAPACITIES

1. Complete the first water budget below for Houston, Texas (Table 9.5a). Notice differences occur by changing the soil texture. The first ledger is for fine sand, and the second is for silty loam. Start your calculations in the month following the gray background ST value.

TABLE 9.5 TWO SOIL MOISTURE BUDGETS FOR HOUSTON, TEXAS

(a) Fine Sand

30°N, 95°W — Elevation 19 m — Field capacity = 102 mm

	JAN	FEB	MAR	APR	MAY	JUN	JUL	AUG	SEP	OCT	NOV	DEC	YR
P	103	89	78	89	133	131	105	118	130	93	124	93	1286
PE	20	27	46	76	109	146	163	164	131	88	46	25	1041
P-PE	+83	+62	+32	+13	+24		−58	−46	−1		+78	+68	
DST	0	0	0	0	0	−58			0		+78	+19	
ST	102	102	102	102	102				0		83		
AE	20	27	46	76	109	146	163	147	130	88	46	25	1023
D	0	0	0	0	0	0			1	0	0		18
S	83	62	32	13	24	0	0	0	0	0	0	49	263

(b) Silty Loam

30°N, 95°W — Elevation 19 m — Field capacity = 203 mm

	JAN	FEB	MAR	APR	MAY	JUN	JUL	AUG	SEP	OCT	NOV	DEC	YR
P	103	89	78	89	133	131	105	118	130	93	124	93	1286
PE	20	27	46	76	109	146	163	164	131	88	46	25	1041
P-PE	+83	+62	+32	+13	+24	−15	−58	−46	−1	+5	+78	+68	
DST	0	0	0	0	0	−15	−58	−46	−1	+5	+78	+37	
ST	203	203	203	203	203	188	130	84	83	88	166	203	
AE	20	27	46	76	109	146	163	164	131	88	46	25	1041
D	0	0	0	0	0	0	0	0	0	0	0	0	0
S	83	62	32	13	24	0	0	0	0	0	0	31	245

105

2. a. Silty loam is a finer texture than fine sand. What effect does finer soil texture have on annual deficit in Houston?

   b. Why?

3. a. Which Houston soil texture should experience lower stream stage levels, lake levels, and water tables, and probably somewhat greater forest fire risk in summer (April–September)?

   b. How did you know which soil texture to pick?

4. a. What effect does finer soil texture have on the amount of annual AE in Houston?

   b. Why?

5. a. Evapotranspiration includes water yielded to the atmosphere by plant growth during the summertime (April–September growing season). Assuming that the plant-generated proportion of AE is constant, which Houston soil texture should have slightly better vegetation growth?

   b. Why?

6. a. What effect does finer soil texture have on the amount of annual surplus in Houston?

   b. Why?

7. Why are coarse-textured soils more hydrologically responsive (flood and dry more rapidly) than fine-textured soils?

Name: _____ Section: _____

PART 3 • LATITUDE, ALTITUDE, AND CONTINENTALITY

Washington, D. C. (Table 9.6), and Boulder, Colorado (Table 9.7), are at the same latitude, but they have very different water budgets. Notice, however, that two other dimensions of environmental variability are also very different between these cities. Boulder is well over a kilometer higher in altitude than Washington, D. C. Boulder also is more continental, being far in the interior of North America and thus well removed from any oceans.

TABLE 9.6 SOIL MOISTURE BUDGET FOR WASHINGTON, D. C.

39°N, 77°W	Elevation 3 m				Fine sand			Field capacity = 102 mm				
	JAN	FEB	MAR	APR	MAY	JUN	JUL	AUG	SEP	OCT	NOV	DEC	YR
P	70	66	90	72	94	80	97	104	85	78	76	79	991
PE	1	4	22	47	80	116	135	129	99	56	28	7	724
P-PE	+69	+62	+68	+25	+14	−36	−38	−25	−14	+22	+48	+72	
ΔST	0	0	0	0	0	−36	−38	−25	−3	+22	+48	+32	
ST	102	102	102	102	102	66	28	3	0	22	70	102	
AE	1	4	22	47	80	116	135	129	88	56	28	7	713
D	0	0	0	0	0	0	0	0	11	0	0	0	11
S	69	62	68	25	14	0	0	0	0	0	0	40	278

1. a. Between these two sites—Washington, D. C., and Boulder, Colorado—what variable (P, PE, AE, D, or S) has the greatest difference in annual total?

   b. Why?

2. Since Boulder does get precipitation, why is there rarely any surplus?

3. a. How much less is annual PE in Boulder than in Washington, D.C.?

107

TABLE 9.7 SOIL MOISTURE BUDGET FOR BOULDER, COLORADO

40°N,105°W		Elevation 1638 m				Fine sand			Field capacity = 102 mm				
	JAN	FEB	MAR	APR	MAY	JUN	JUL	AUG	SEP	OCT	NOV	DEC	YR
P	14	20	40	65	78	48	47	38	39	38	25	29	481
PE	0	3	14	36	61	90	109	104	78	47	17	2	561
P-PE	+14	+17	+26	+29	+17	−42	−62	−66	−39	−9	+8	+27	
ΔST	+14	+17	+26	+18	0	−42	−60	0	0	0	+8	+27	
ST	49	66	92	102	102	60	0	0	0	0	8	35	
AE	0	3	14	36	61	90	107	38	39	38	17	2	445
D	0	0	0	0	0	0	2	66	39	9	0	0	116
S	0	0	0	19	17	0	0	0	0	0	0	0	36

   b. Why is PE consistently lower in Boulder than in Washington, D.C.?

4. Why is AE consistently lower in Boulder than in Washington?

5. Houston (fine sand) (Table 9.5a) and Vancouver (Table 9.3) are both coastal cities at low elevation, but sixteen degrees of latitude separate them.

   a. Between these two sites, what variable (P, PE, AE, D, or S) has the greatest difference in annual total?

   b. How much less is annual AE in Vancouver than in Houston?

   c. Comparing Houston to Vancouver, why does AE get lower with higher latitude?

6. Vancouver (Table 9.3) and Stevens Point (Table 9.4) are at about the same latitude, but differ by altitude and continentality. Comparing these two sites, annual S differs greatly. Why?

Name: _____   Section: _____

PART 4 ♦ MONSOONAL SOIL MOISTURE BUDGETS

1. Complete the water budget table for sandy loam in Kolkata (Calcutta), India (Table 9.8). Start all calculations in the month following the gray background ST.

TABLE 9.8 — SOIL MOISTURE BUDGET FOR KOLKATA, INDIA

23°N, 88°E		Elevation 2127 m				Sandy loam			Field capacity = _____ mm				
	JAN	FEB	MAR	APR	MAY	JUN	JUL	AUG	SEP	OCT	NOV	DEC	YR
P	12	25	32	53	129	291	329	338	266	131	21	7	
PE	44	73	152	220	234	212	186	182	184	156	89	47	
P-PE													
ΔST													
ST					0								
AE													
D													
S													

2. Draw the water budget graph on Figure 9.4 showing soil moisture seasons for Kolkata.

3. What kind of climate or biome would you say Kolkata has? Why do you think so?

4. What is probably a serious hazard for people living in Kolkata?

109

FIGURE 9.4 SOIL MOISTURE BUDGET GRAPH FOR KOLKATA, INDIA

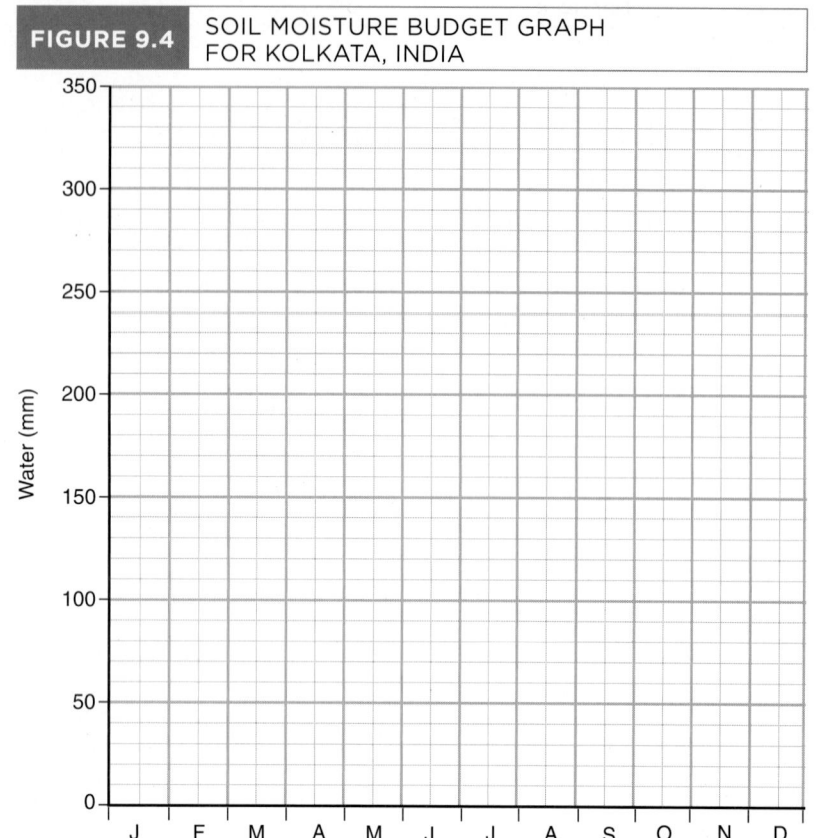

EXERCISE 10 † ANALYSIS OF SOIL MOISTURE PROPERTIES

PURPOSE

The purpose of this exercise is to learn how to measure soil moisture characteristics, and to learn the relationship between soil texture and moisture characteristics through scientific experimentation.

LEARNING OBJECTIVES

By the end of this exercise you should be able to

- use the scientific method to test hypotheses relating soil texture to soil moisture characteristics;
- explain the relationship between soil texture and field capacity;
- explain the relationship between soil texture and infiltration capacity;
- explain how the infiltration rate changes over time;
- distinguish field capacity from infiltration capacity;
- distinguish gravity water from capillary water;
- use a soil texture triangle to classify soil texture; and
- explain the relationship between soil texture, wilting point, and available water capacity.

INTRODUCTION

Soil is a mixture of inorganic matter (minerals), organic matter, air, and water that thinly mantles the surface of the earth. Soil forms in response to processes occurring in the atmosphere, hydrosphere, biosphere, and lithosphere. Since the processes operating in these four spheres of the physical environment vary greatly from place to place, soils should also vary from place to place, creating geographic patterns to the earth's physical environment. **Soil properties** are important for describing, differentiating, and classifying soils from one place to another, and for distinguishing different **horizons** (layers) within a soil. The six major properties of soil are color, texture, structure, organic content, moisture, and chemical characteristics. This experiment focuses only on texture and moisture characteristics of soil.

Soil Texture

Soil texture refers to the relative proportions of different-sized particles comprising a soil. There are three main classes of particle sizes: **clay** particles are less than 0.002 millimeters in diameter, **silt** particles range from 0.002 to 0.05 millimeters in diameter, and **sand** particles range from 0.05 to 2.0 millimeters in diameter. Once the relative proportions of clay, silt, and sand are determined, **soil texture class** is identified using a soil texture triangle (Figure 10.1).

Soil texture is important for a variety of reasons. Texture affects the rate and intensity of chemical weathering processes. The finer the texture, the more surface area there is available for chemical reactions to occur on (Figure 10.2). Likewise, soil texture affects the availability of nutrients for plant use. As more surface area is exposed, the availability of nutrients increases. More importantly for this exercise, soil texture affects soil moisture characteristics. Weathering processes, nutrient availability, and plant processes all require moisture.

SOIL MOISTURE

Pore spaces are voids between soil particles that can hold air or water. **Porosity,** the total volume of pore space in a given volume of soil, is determined by soil texture. In general, fine-textured soils, such as clay, have a higher porosity than coarse-textured soils, such as sand. Soils consisting of uniformly sized particles have a higher porosity than soils consisting of a mixture of particle sizes. **Permeability** is the relative ease with which water can move through a soil, and is determined by how well individual pore spaces are connected. Although clay soils have a higher porosity than sandy soils, they have a much lower permeability; water cannot move through clay soils as easily as through sandy soils. The pores in sandy soils tend to be very well connected and water moves rapidly through sand. The opposite is true for clay.

The porosity and permeability of a soil determine the amount of water and the rate at which that water may soak into the soil. The **infiltration rate** is the rate at which water soaks into the soil. The infiltration rate for a given soil texture may vary depending on the amount of moisture already in the soil and on the local land use conditions, such as vegetation cover. The **infiltration capacity** is the maximum *steady* rate at which water can infiltrate a soil.

Once water has infiltrated the soil, it may flow downward under the influence of gravity to become part of the regional groundwater flow, or it may flow laterally through the soil. **Gravity water** is water that quickly drains downward through the pore spaces in the soil under the influence of gravity. Gravity water is not available for use by plants, but plays an important role in the downward movement of fine soil particles, nutrients, minerals, and chemicals such as pesticides, insecticides, and other water pollutants. Gravity water, therefore, plays an

FIGURE 10.1 SOIL TEXTURE TRIANGLE

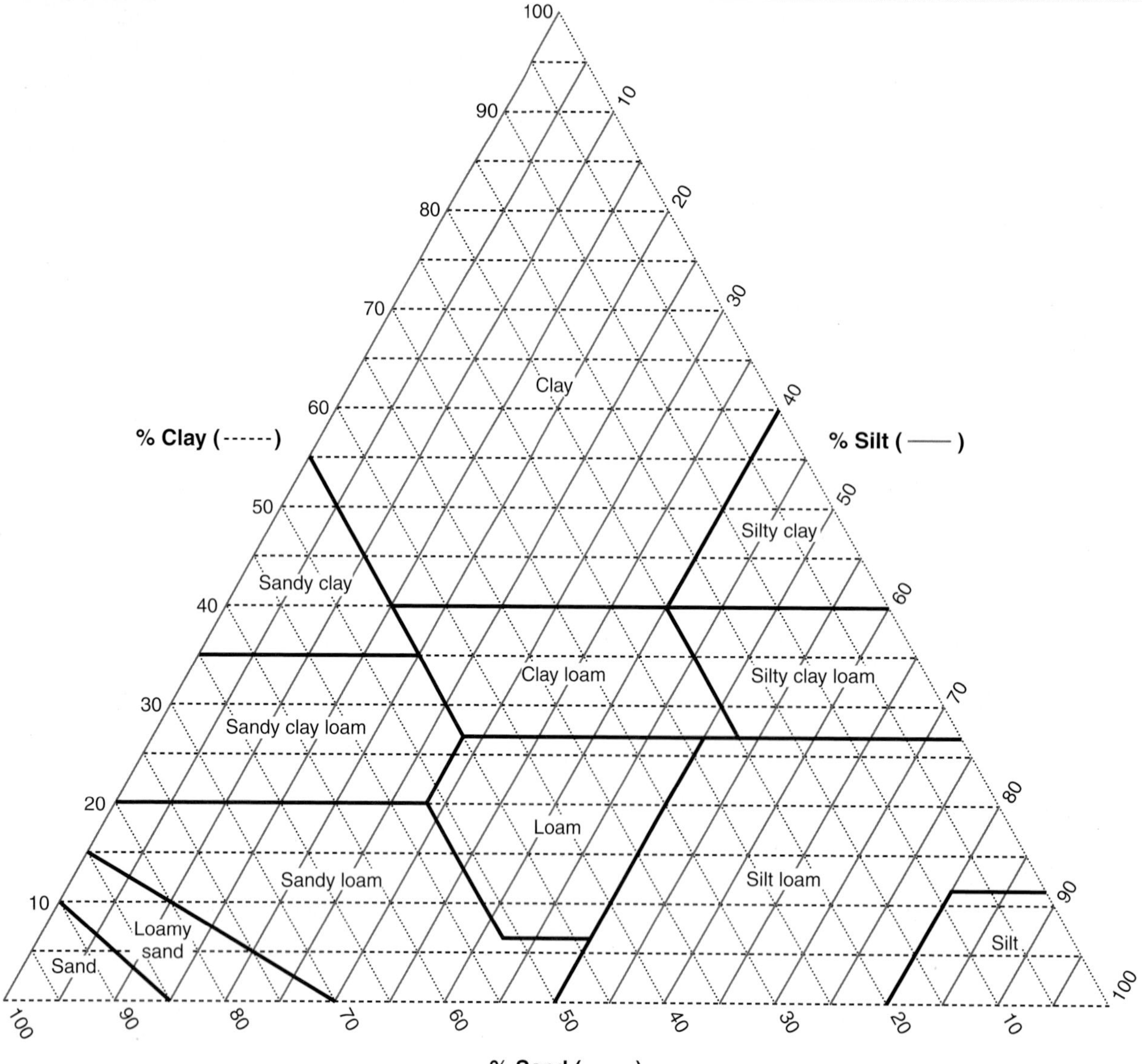

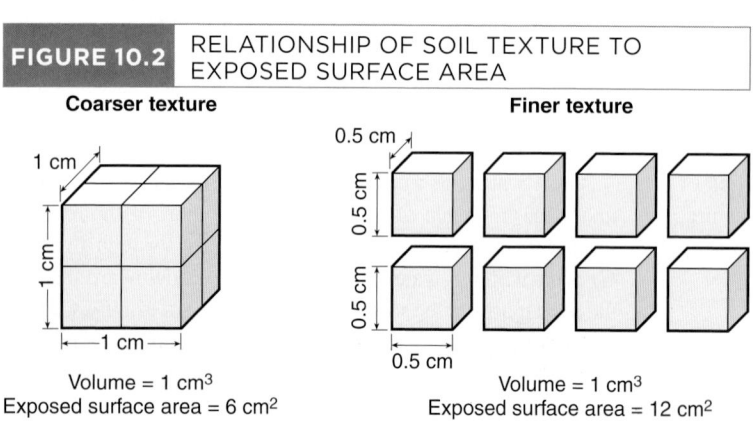

Source: Data from USDA, National Soil Survey Handbook.

FIGURE 10.2 RELATIONSHIP OF SOIL TEXTURE TO EXPOSED SURFACE AREA

Coarser texture

Volume = 1 cm³
Exposed surface area = 6 cm²

Finer texture

Volume = 1 cm³
Exposed surface area = 12 cm²

important role in determining the quality of groundwater, an important source of drinking water in some places.

Some water that infiltrates the soil may be stored there. This stored water may be used by plants, or may be evaporated. **Capillary water** is water that remains in the soil after the gravity water has drained away. This is the principle source of water for plant use. **Field capacity** is the upper limit of soil moisture after gravity water has drained away; it is the maximum amount of water the soil can hold. As plants access capillary water, the soil slowly dries until it reaches the **wilting point,** the point at which soil water is no longer available to plants. The **available water capacity** is the difference between field capacity and the wilting point, and this is the amount of water actually available for use by plants. Hence, available water capacity affects how well plants grow in different soils.

Once the wilting point is reached, the soil is still not totally dry. Hygroscopic water still remains. **Hygroscopic water** is water bound to soil particles so tightly that it is not available for plant use. The field capacity, wilting point, and available water capacity of a soil all depend on soil texture (Figure 10.3).

Scientific Hypotheses

The **scientific method** constitutes a process of controlled inquiry and it involves following a particular procedure to answer questions or test hypotheses about the world around us. In other words, science involves experimentation. Science also involves making careful observations and measurements during the course of experiments. Valid scientific experiments must be repeatable, and the outcomes must be verifiable.

Scientific experiments often start with a **hypothesis,** a statement that scientists believe to be true, and that is testable. In order to test a hypothesis, scientists must first define a procedure to follow, and then they must follow it carefully. If the procedure is unclear or if mistakes are made during the procedure, the outcome of the experiment may be incorrect or may be unverifiable by other scientists. In this exercise, therefore, it is important to follow carefully the procedures outlined, otherwise the outcome of the experiment may be questionable.

In any experiment, no matter how careful scientists are, they are still human and occasionally make mistakes. This can result in unexpected results, and it is one of the reasons why scientists insist on experiments being repeated numerous times before placing confidence in the outcomes of their experiments. In general, if an experiment is repeated numerous times, and a large percentage of the results agree with the hypothesis, then scientists will accept the hypothesis as being true despite a few outcomes that may suggest otherwise. It is important that any outcomes in disagreement with the hypothesis be a result of random error rather than actual differences in the thing being tested. In other words, scientists don't reject a hypothesis just because one outcome out of many disagrees with the hypothesis. At the same time, if numerous outcomes disagree with the hypothesis, then scientists should either reconsider their hypothesis, or they should reconsider the procedures they are following to test the hypothesis.

In this exercise, three hypotheses regarding soil texture and soil moisture will be tested:

Hypothesis 1: Coarse-textured soils have lower field capacities than fine-textured soils.

Hypothesis 2: For all soil textures, the infiltration rate is high for dry soils, and as soil moisture increases, the infiltration rate decreases.

Hypothesis 3: Coarse-textured soils have higher infiltration capacities than fine-textured soils.

IMPORTANT TERMS, PHRASES, AND CONCEPTS

soil	infiltration capacity
soil properties	gravity water
horizons	capillary water
soil texture	field capacity
clay, silt, sand	wilting point
soil texture class	available water capacity
pore space	hygroscopic water
porosity	scientific method
permeability	hypothesis
infiltration rate	

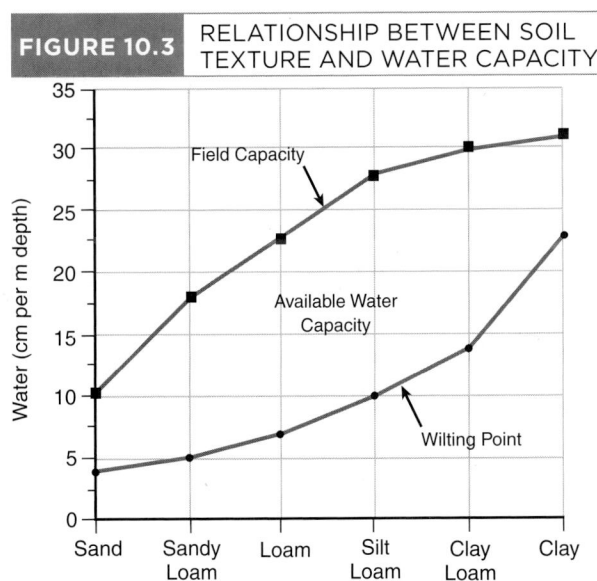

FIGURE 10.3 RELATIONSHIP BETWEEN SOIL TEXTURE AND WATER CAPACITY

Source: Data from Steila et al., 1989.

REFERENCES

U.S. Department of Agriculture, Natural Resources Conservation Service. 2007. *National Soil Survey Handbook, title 430-VI.* [online]

Steila, D. and T. E. Pond. 1989. *The Geography of Soils: Formation, Distribution, and Management.* 2d ed. Maryland: Rowman & Littlefield.

PROCEDURE: SOIL MOISTURE ANALYSIS

1. Record in Table 10.1 the different soils and their textures. This information will be provided by your instructor.
2. Record in Table 10.2 which soil sample you are analyzing and the weight of this sample.
3. Fill a graduated cylinder to the 50 ml mark. Slowly pour the 50 ml of water onto the soil. Use a stop watch to time how long it takes for this water to completely soak into the soil. Start timing when you start pouring water on the sample and stop timing when there is no water left standing on top of the soil. Record this time in Table 10.2 under "Pour 1." Water may still be coming out the bottom of the sample, but this is not what we're timing. Only time how long it takes for water to soak into the top of the soil.
4. Some of this water will filter out the bottom of the sample into the graduated cylinder. This is gravity water. Record in Table 10.2 under "Amount water in cylinder" the amount of water which drips out of the soil and into the graduated cylinder.
5. The amount of water left in the soil equals the difference between what was poured into the soil (50 ml) and what dripped out the bottom:

$$\text{amount water left in soil} = 50 - \text{amount water in cylinder}.$$

   Enter this amount in Table 10.2 under "Amount water left in soil."
6. The "New soil moisture 1" in Table 10.2 equals the "Starting soil moisture" plus the "Amount water left in soil" (out of this 50 ml).
7. Empty the graduated cylinder under the sample containing gravity water.
8. Carefully pour another 50 ml of water into the soil container and time how long it takes to completely infiltrate the soil. Record this time in Table 10.2 under "Pour 2."
9. Record the amount of gravity water that drips into the graduated cylinder.
10. The "Amount water left in soil" from this pour is the difference between what was poured in this time (50 ml) and the water that dripped out the bottom:

$$\text{amount water left in soil} = 50 - \text{amount water in cylinder}.$$

    Enter this amount in Table 10.2 under "Amount water left in soil."
11. "New soil moisture 2" equals the "Amount water left in soil" from this pour plus "New soil moisture 1."
12. Empty the graduated cylinder under the sample containing gravity water.
13. Repeat steps (8) through (12) until the "New soil moisture" does not change for at least three consecutive pours and the "Time to infiltrate" also does not change for at least three consecutive pours (or until your instructor tells you to stop).

Name: _____   Section: _____

PART 1 | DATA COLLECTION

TABLE 10.1 SUMMARY OF ALL SOIL TEXTURES (PERCENTAGES)

Texture Class	Soil A	Soil B	Soil C	Soil D	Soil E	Soil F	Soil G
Very coarse sand (1.0–2.0 mm)							
Coarse sand (0.5–0.99 mm)							
Medium sand (0.25–0.49 mm)							
Fine sand (0.125–0.249 mm)							
Very fine sand (0.0625–0.1249 mm)							
Silt (<0.0625 mm)							

TABLE 10.2 SOIL MOISTURE DATA

Which soil are you analyzing? _____   Soil weight: _____
Starting soil moisture (ml)   _____ 0 ml _____

Pour 1 Time to infiltrate 50 ml (min:sec) _____
    Amount water in cylinder (ml) _____
    Amount water left in soil (out of this 50 ml)   _____
    New soil moisture 1 (ml)   _____

Pour 2 Time to infiltrate 50 ml (min:sec) _____
    Amount water in cylinder (ml) _____
    Amount water left in soil (out of this 50 ml)   _____
    New soil moisture 2 (ml)   _____

Pour 3 Time to infiltrate 50 ml (min:sec) _____
    Amount water in cylinder (ml) _____
    Amount water left in soil (out of this 50 ml)   _____
    New soil moisture 3 (ml)   _____

Pour 4 Time to infiltrate 50 ml (min:sec) _____
    Amount water in cylinder (ml) _____
    Amount water left in soil (out of this 50 ml)   _____
    New soil moisture 4 (ml)   _____

(continued)

TABLE 10.2	SOIL MOISTURE DATA *(CONTINUED)*

Pour 5 Time to infiltrate 50 ml (min:sec) _____

    Amount water in cylinder (ml) _____

    Amount water left in soil (out of this 50 ml)   _____

    New soil moisture 5 (ml)   _____

Pour 6 Time to infiltrate 50 ml (min:sec) _____

    Amount water in cylinder (ml) _____

    Amount water left in soil (out of this 50 ml)   _____

    New soil moisture 6 (ml)   _____

Pour 7 Time to infiltrate 50 ml (min:sec) _____

    Amount water in cylinder (ml) _____

    Amount water left in soil (out of this 50 ml)   _____

    New soil moisture 7 (ml)   _____

Pour 8 Time to infiltrate 50 ml (min:sec) _____

    Amount water in cylinder (ml) _____

    Amount water left in soil (out of this 50 ml)   _____

    New soil moisture 8 (ml)   _____

Pour 9 Time to infiltrate 50 ml (min:sec) _____

    Amount water in cylinder (ml) _____

    Amount water left in soil (out of this 50 ml)   _____

    New soil moisture 9 (ml)   _____

Name: _____   Section: _____

PART 2 · ANALYSIS AND INTERPRETATION

1. To test our hypotheses, we must rank the soils according to texture. Using the information presented in Table 10.1, rank the soils from coarsest to finest.

   coarsest: _____

   _____

   _____

   _____

   _____

   _____

   finest: _____

2. What problems, if any, did you encounter analyzing the soil moisture characteristics of your soil sample? Problems could include timing, reading water quantities, measuring water quantity, misrecording information, etc. Think of all the errors you or your peers may have encountered while measuring moisture characteristics. These errors may affect the success of this experiment.

3. Determine the field capacity for a 100 gram sample of your soil. The last "New soil moisture" you recorded in Table 10.2 equals the field capacity for your entire sample. If your soil sample weighed 100 grams, then your last soil moisture equals the field capacity, and this should be recorded under (b) below (ignore step a). If your sample weighed more than 100 grams, follow steps (a) and (b).

   a. Divide the weight of the soil sample by 100.   _____

   b. Divide the last "New soil moisture" by your answer to part (a).

   This is the field capacity for 100 grams of soil.   _____

4. Collect information from your peers on the field capacity for 100 grams of their soil. If there are multiple samples of the same soil, average the field capacities and record this information in Table 10.3.

| TABLE 10.3 | FIELD CAPACITY SUMMARY (ML WATER PER 100 G SOIL) |

	Soil A	Soil B	Soil C	Soil D	Soil E	Soil F	Soil G
Field capacity							

5. To test hypothesis 1, we must rank the soils according to field capacity. Using the information in Table 10.3, rank the soils from lowest to highest field capacity.

   lowest field capacity: _____

   _____

   _____

   _____

   _____

   _____

   highest field capacity _____

6. If hypothesis 1 is correct, the soil with the lowest field capacity should be the soil with the coarsest texture, and the soil with the highest field capacity should be the soil with the finest texture. Compare your ranking of soils according to field capacity to your ranking of soils according to texture. The order of the ranking is the important point here, not the absolute position of a particular soil sample.

   a. Which soil samples are in the right order?

   b. Which soil samples are in the wrong order?

7. On the basis of these results, should we accept hypothesis 1 as true, should we reject it as false, or should we accept it conditionally (assume it is partially true)? Why? Remember, mismatches could be a result of problems encountered while collecting our data (see question 2).

8. The infiltration rate is the time it takes for a certain amount of water to soak into the soil. Determine the infiltration rate for each pour, following the procedure outlined here.

   a. Copy, from Table 10.2 to column 1 of Table 10.4, the time it took for all the water to infiltrate the soil for each pour. These times should be in minutes and seconds.

   b. Convert these times to seconds only (1 minute = 60 seconds) and record the time in seconds only in column 2 of Table 10.4.

   c. Determine the infiltration rate by dividing the amount of water poured into the soil (50 ml) by the time it took (in seconds) to completely soak in:

   $$\text{Rate} = \frac{50 \text{ ml}}{\text{time to infiltrate (seconds)}}$$

   Record this infiltration rate in column 3 of Table 10.4.

   d. Multiply the infiltration rate in column 3 by 60 to convert the infiltration rate to milliliters per minute. Record this final rate in column 4.

9. To test hypothesis 2, we must compare the infiltration rates for the different soils. Collect information from your peers on the infiltration rate (in ml/min) for their soils. If there are multiple samples of the same soil, average the rates for each pour and record this information in Table 10.5.

10. Draw a graph of the infiltration rates in Table 10.5. Plot the infiltration rate for each soil as a smooth line on the graph paper provided. The infiltration rate should be graphed on the Y-axis and the pours should be graphed on the X-axis. You will have one line on your graph for each soil. Label these lines clearly.

TABLE 10.4 CALCULATION OF INFILTRATION RATE (ML/MIN)

	Time to infiltrate (min:sec)	Time to infiltrate (sec)	Infiltration rate (ml/sec)	Infiltration rate (ml/min)
Pour 1				
Pour 2				
Pour 3				
Pour 4				
Pour 5				
Pour 6				
Pour 7				
Pour 8				
Pour 9				

TABLE 10.5 SUMMARY OF INFILTRATION RATES (ML/MIN)

	Soil A	Soil B	Soil C	Soil D	Soil E	Soil F	Soil G
Pour 1							
Pour 2							
Pour 3							
Pour 4							
Pour 5							
Pour 6							
Pour 7							
Pour 8							
Pour 9							

11. Hypothesis 2 states that the infiltration rate is high for dry soils and decreases as soil moisture increases.

    a. On the basis of the trends shown in the graph you drew for question 10, should we accept this hypothesis as true, as partially true, or as false?

    b. Why? Remember, problems encountered during the experiment may affect the outcome.

    c. If the results do not support the hypothesis, devise a new hypothesis that reflects what the results show. This hypothesis should be a statement that is unambiguous, that appears to be true, and that is testable.

12. The infiltration capacity is the maximum steady rate at which water can soak into the soil. Your graphs from question 10 should show that infiltration rates start out high, decrease, and eventually level off. The rate at which infiltration levels off is considered the *infiltration capacity* for that soil. Using your graph, determine the infiltration capacity for each soil.

    Soil A: _____

    Soil B: _____

    Soil C: _____

    Soil D: _____

    Soil E: _____

    Soil F: _____

    Soil G: _____

13. To test hypothesis 3, we must rank the soils by infiltration capacity. Rank the soils from highest to lowest infiltration capacity.

    highest infiltration capacity: _____

    _____

    _____

    _____

    _____

    _____

    lowest infiltration capacity: _____

14. If hypothesis 3 is correct, the soil with the highest infiltration capacity should be the soil with the coarsest texture, and the soil with the lowest infiltration capacity should be the soil with the finest texture. Compare your ranking of soils according to infiltration capacity to your ranking of soils according to texture. The order of the soils is the important point, not the absolute position of each sample.

    a. Which soil samples are in the correct order?

    b. Which soil samples are not in the correct order?

15. On the basis of these results, should we accept hypothesis 3 as true, should we reject it as false, or should we accept it conditionally (assume it is partially true)? Why? Remember, mismatches could be a result of problems encountered while collecting our data.

16. What conclusions can you make regarding the relationship between soil texture and moisture characteristics?

17. a. If you did not accept each of the hypotheses as true, or if your results did not completely support the hypotheses, was this because the hypotheses were really false, or was it because you or your peers may have made mistakes or errors during the experiment?

    b. If you wanted to repeat this experiment with the hope of getting results that were more consistent with our hypotheses, how might you set up the experiment differently to reduce the chance of errors?

Name: _____   Section: _____

PART 3 † THE SOIL TEXTURE TRIANGLE

1. Soil texture is usually defined based on the percent clay, silt, and sand, yet the samples we analyzed in parts 1 and 2 contain only gradations of sand (coarse, medium, fine), and the finest particle size is silt. Analysis of some other soil samples reveals the following information on particle size. Use the soil texture triangle (Figure 10.1) to determine the texture class of these soil samples. Then use Figure 10.3, which relates texture class and soil moisture, to determine the field capacity, wilting point, and available water capacity of each soil. Record this information in Table 10.6.

TABLE 10.6 SOIL TEXTURE AND MOISTURE PROPERTIES

	% Sand	% Silt	% Clay	Texture Class	Field Capacity	Wilting Point	Available Water
Soil T	10	20	70				
Soil V	90	5	5				
Soil W	60	25	15				
Soil X	40	40	20				
Soil Y	30	40	30				
Soil Z	20	70	10				

2. Of the soils listed in Table 10.6, which soil texture class has the highest available water capacity?

3. Of the soils listed in Table 10.6, which two soil texture classes have the lowest available water capacity?

4. Explain why soil texture and available water capacity are related in the manner you described in questions 2 and 3.

EXERCISE 11 · CLIMATE, NET PRIMARY PRODUCTION, AND DECOMPOSITION

PURPOSE

The purpose of this exercise is to demonstrate relationships between climate, vegetation growth, and rates of organic decomposition in North America.

LEARNING OBJECTIVES

By the end of this exercise you should be able to

- describe geographic patterns of net primary production in North America;
- describe geographic patterns of decomposition rates in North America;
- explain how actual evapotranspiration affects net primary production and how it affects decomposition rates;
- explain how net primary production values correspond to different types of biomes; and
- explain how climatic conditions and physiographic characteristics may result in steep net primary production gradients.

INTRODUCTION

Net primary production (NPP) is the normal amount of new living plant tissue gained by growth, less the amount lost through death or relocation, per unit area, per unit interval of time. In this exercise NPP is measured as the total weight of new material in all living plants, called plant **biomass,** produced annually within a square meter area over the course of a year ($g/m^2/yr$). NPP is an important index of how well vegetation lives within the ecological community at a place.

Only expensive and painstakingly tedious methods can produce a direct *measurement* of plant biomass. Essentially, every part of every single plant (roots, leaves, stems, etc.) must be separated from all non-plant material in a test area, and weighed. Aside from the drawback of being labor-intensive, this inherently destructive testing is unacceptably wasteful. It also precludes any before-and-after determination of biomass changes. Therefore, scientists use indirect means to *estimate* NPP, but these estimates must be based in observable relationships from those few places where measurements of both biomass and the variables controlling plant growth do exist. Various predictors are conceivable, but a twofold problem arises: (1) determining what variables are genuine controls over plant growth, and then (2) either finding, gathering, or calculating representative numerical data for these control variables.

Estimating NPP with AE

Plant growth is especially sensitive to two climatic variables, temperature and moisture, both of which affect evapotranspiration. Potential evapotranspiration (PE) is the moisture that could be yielded to the atmosphere if the water supply was limitless, and is solely a function of temperature. **Actual evapotranspiration** (AE) represents the amount of moisture actually yielded to the atmosphere, and is limited by *both* temperature and moisture availability. If conditions are *both* hot and wet, AE will be high. If conditions are *either* cold *or* dry, AE will be low. Thus, to plants actual evapotranspiration (AE) represents the true utility of water—its availability in necessary quantity and quality, at the correct phase, and during the growing season. Because of this, AE provides a more realistic single expression of climate than temperature, PE, or precipitation alone when specifying the relationship between climate and biosphere production. AE values can be calculated for any place where we have climate data from a water budget. Since AE exerts a genuine control over plant growth, AE may be used to estimate NPP. In this exercise we examine the spatial variability of plant production across North America, as indexed by net primary production (NPP), and predicted solely by AE.

Figure 11.1 shows an example where scientists from the Environmental Protection Agency gathered numerical data for actual evapotranspiration and net primary production in Big Cypress Swamp, Florida. The trend line on the graph in Figure 11.1 shows the relationship between AE and NPP. This trend line shows that as AE increases, NPP also increases.

Graphic estimation of NPP for any site can be accomplished using the trend line. First find the AE value of the site along the *X*-axis. Second, use a straightedge to draw a vertical line upward from the AE point on the *X*-axis to wherever the straightedge intersects the trend line. Third, draw a horizontal line from the trend line intersection over to the *Y*-axis. The value at the *Y*-axis is the estimate of NPP for that site.

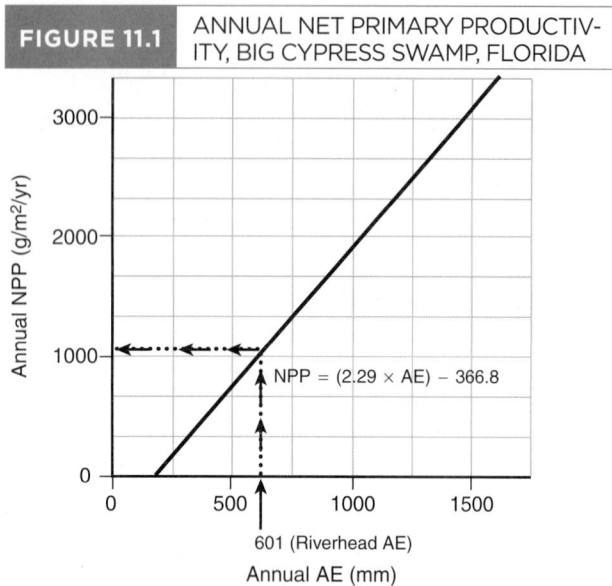

FIGURE 11.1 ANNUAL NET PRIMARY PRODUCTIVITY, BIG CYPRESS SWAMP, FLORIDA

Source: Adapted from U.S. EPA, 1973.

The example shown in Figure 11.1 is for Riverhead, New York, which has an annual AE of 601 mm. The estimated NPP is about 1000 g/m^2/yr, meaning that on average each square meter of ground in Riverhead gains a kilogram (about 2.4 lbs) of new plant matter each year.

Calculated estimation of NPP for any site uses an equation defining the graph trend line. Calculated estimates may be more desirable than graphic estimates because the graph is not as accurate as an equation, and you might not always have a copy of this graph with you. The equation defining the AE:NPP relationship shown by the trend line is:

$$\text{NPP} = (2.29\text{AE}) - 366.8. \quad (11.1)$$

The calculation approach to estimating NPP enables rapid computation of much more precise NPP values. For Riverhead, New York, this equation gives an estimate of 1010 g/m^2/yr.

Production Gradients

NPP differs at separate places, but patterns in the rate of NPP change between places are also revealing. A **production gradient** expresses the change in production over distance:

$$\text{Production gradient} = \frac{\Delta \text{NPP}}{\text{Distance}} \quad (11.2)$$

where Δ (delta) stands for "change in."

In some regions, or some types of habitats, the rate of change in NPP is so gradual as to be almost imperceptible even after crossing large ground distances. In other places, however, NPP is radically different between places only a short distance apart. The latter example falls along a steep (large) production gradient, whereas the first illustrates a flat or shallow (small) production gradient.

Steep production gradients result from dramatic changes to the conditions that control plant growth over short horizontal distances. Often these conditions are climatic, and in turn the climatic conditions may be influenced by other environmental attributes. Mountainous terrain often exhibits dramatic climatic contrasts over short distances, and thus helps create some steep production gradients. Not all steep production gradients arise from topography and climate, however. Steep local gradients in NPP also may arise from geographic variability in things such as soil nutrients/toxins, seaside salinity, and ground permeability.

Estimating Decomposition Rates with AE

In genuine ecological communities there is constant recycling of nutrients from the decomposition of **litter** (dead organic material), through the soil, to newly produced living biomass. The potential **decomposition rate** (and thus nutrient recycling rate) may also be estimated using a known relationship with AE. Figure 11.2 shows this relationship graphically, with annual AE as the X-axis variable and the decomposition rate as the Y-axis variable. Figure 11.2 indicates that a log percentage of a single year's dead organic material mass will break down in one year. Log expressions enable trend lines that actually curve to appear straight by modifying the scale axes (note the Y-axis spacing in Figure 11.2). This exercise provides all decomposition rates, so do not divert attention to log relationships. The example in Figure 11.2 shows that Riverhead, with an AE of 601mm, has a decomposition rate of 60% (0.60).

The higher the decomposition rate, the more vigorous is chemical reactivity, and the more rapid the recycling rate of dead organic material. At exactly 100%, all litter production for each year decomposes within exactly one year. At 1900 mm of annual AE, the decomposition rate is 112%, which

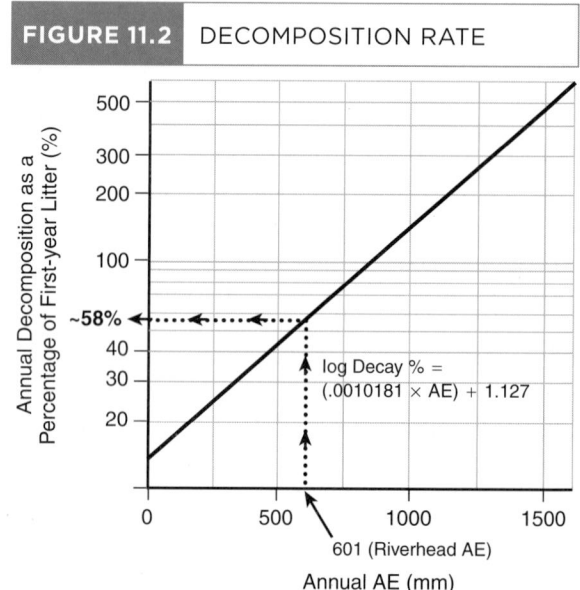

FIGURE 11.2 DECOMPOSITION RATE

Source: Adapted from Meentemeyer et al., 1977.

means that all organic litter produced at such a site each year could decompose in less than a year. In such a place, the ground would be almost devoid of litter accumulation. Decay rates exceeding 100% occur only in exceptionally hot and wet conditions.

On the other hand, at sites where the annual decomposition rate is less than 100%, litter accumulates on the ground because it takes longer to decompose than to produce. At Riverhead, for example, only 60% of each year's litter decomposes within a typical year, so it takes almost two years to decompose a single year's litter production. If this happens at sites that are at least seasonally dry, the net litter accumulates through time and represents a growing *fuel load*, which eventually may decompose in a far more catastrophic fashion—wildfire. In fact, in many low AE habitats, wildfire serves the useful function of episodically recycling nutrients, thus compensating for the naturally low decomposition rate. Fire can become a normal recurrence, to which ecological communities must adapt.

Decay time is the number of years it takes to decompose all of one year's litter. This is sometimes a more imaginable index of how rapidly nutrient recycling occurs at a place. Decay time is calculated by:

$$\frac{1}{X} \quad (11.3)$$

where X is the decomposition rate expressed as a decimal ratio (e.g., 60% = 0.60).

Riverhead's 60% decomposition rate (0.60) translates into:

$$\frac{1}{0.60} = 1.67 \text{ years}$$

meaning the decay time is one year and eight months, or that every three years Riverhead accumulates two years of undecomposed litter as fuel load. Without fire or other catastrophic recycling, Riverhead accumulates over 10 kilograms of litter on each square meter of ground every fifteen years—clear it, or hope that it stays wet if brush fires are a concern.

IMPORTANT TERMS, PHRASES, AND CONCEPTS

net primary production (NPP)
biomass
actual evapotranspiration (AE)
production gradient

litter
decomposition rate
decay time

REFERENCES

Meentemeyer, et al. 1977. The potential implementation of biogeochemical cycles in biogeography. *The Professional Geographer.* 29: 266–271.

U.S. Environmental Protection Agency, Region IV. 1973. *Ecosystems Analysis of the Big Cypress Swamp and Estuaries.* EPA 904/9-74-002. Washington, D. C.: U.S. Government Printing Office.

Name: _____   Section: _____

PART 1 • ESTIMATING NPP AND DECOMPOSITION FROM AE

1. Annual AE for Green Bay, Wisconsin is 504 mm for fine sand soil.

    a. Using the Figure 11.1 graph, what is Green Bay's NPP? _____ g/m²/yr

    b. Using equation 11.1, what is Green Bay's NPP? _____ g/m²/yr

2. a. Complete the NPP column for Table 11.1 below. Use equation 11.1 and round your answer to the nearest whole number.

    b. Use equation 11.3 to calculate the decay time to the nearest 0.01 year for each site.

TABLE 11.1   SELECTED NORTH AMERICAN SITES

Site	Location	Annual AE (mm)	Annual NPP (g/m²/yr)	Decomposition Rate (%/yr)	Decay Time (yr)
Green Bay, Wisconsin	45°N, 88°W	504		49	
Dryden, Ontario	50°N, 93°W	424		44	
New Orleans, Louisiana	30°N, 90°W	1008		77	
Lamar, Colorado	38°N, 103°W	374		41	
Painted Desert, Arizona	36°N, 112°W	240		32	

3. Some tropical rainforests have an annual AE of 1900 mm.

    What would be their estimated NPP? _____ g/m²/yr

4. Some deserts have an annual AE of only 80 mm.

    What would be their estimated NPP? _____ g/m²/yr

5. Some polar areas have an annual AE of 150 mm.

    What would be their estimated NPP? _____ g/m²/yr

6. The AE:NPP relationship in Figure 11.1 and equation 11.1 derives from direct measurements. Negative NPP values mean that a net loss of vegetation should normally occur every year. If so, no vegetation should exist at all after only a few years, yet there is vegetation in most deserts and polar areas. How do you explain the apparent misprediction? (Hint: Where was the source of Figure 11.1's data?)

7. a. At which Table 11.1 site should a kilogram of leaf litter decompose most rapidly?

   b. Why?

8. Although decomposition values in Table 11.1 might suggest otherwise, Dryden, Ontario, experiences the most rapid accumulation of fuel load. Why? (Hint: Consult Appendix E, Figure E.3 and Appendix F, Figure F.3.)

9. Of the five sites in Table 11.1, where should soil nutrients most accumulate due to lack of consumers (plants)?

10. Considering the predictor variable AE, name two climatic conditions that seem to inhibit decomposition, and indicate an example location (e.g., Amazon rainforest) of where each inhibition factor is likely to prevail.

    Condition _____   Location _____

    Condition _____   Location _____

Name: _____   **Section:** _____

PART 2 ✟ MAPPING AND MAP INTERPRETATION OF NPP AND DECOMPOSITION

1. NPP values from Table 11.4 are plotted on the North America map in Figure 11.3.

   a. Add the five NPP values you calculated in Table 11.1 to the map blanks.

   b. Draw NPP isolines for 0, 250, 500, 750, 1000, 1500, and 2000 g/m²/yr on the North America map. (Refer to Appendix B for information on drawing isolines.)

2. a. What NPP value corresponds with a 2.5-year decay time (40% decomposition rate)?

   b. In a distinct color, draw approximate boundaries separating the more-than from less-than 2.5-year decay time area on the map.

3. a. In about what fraction (e.g., one-half, one-tenth, etc.) of North America does one full year's dead organic materials decompose in 2.5 years or longer (i.e., the decomposition rate is slow)?

   b. Where do these areas occur?

   c. Using Figure 11.4, determine what climate conditions (temperature and precipitation) occur in these areas.

   d. Compare these areas with the vegetation maps in Appendix E, Figure E.3 and Appendix F, Figure F.3. What kinds of vegetation correspond with slow decomposition?

   e. Describe how these places seem to correspond to areas on the potential wildfire map in Figure 11.4.

4. Your isoline map in Figure 11.3 should depict a large vertical band in western North America that has an NPP below 500 g/m²/yr. Compare this area with the biomes map in Appendix F, Figure F.3. What biome(s) coincide with this low-production zone:

   a. north of about 55°N?

   b. between about 40°N and 55°N?

   c. between about 25°N and 40°N?

FIGURE 11.3 ESTIMATED NORTH AMERICAN NPP (G/M²/YR)

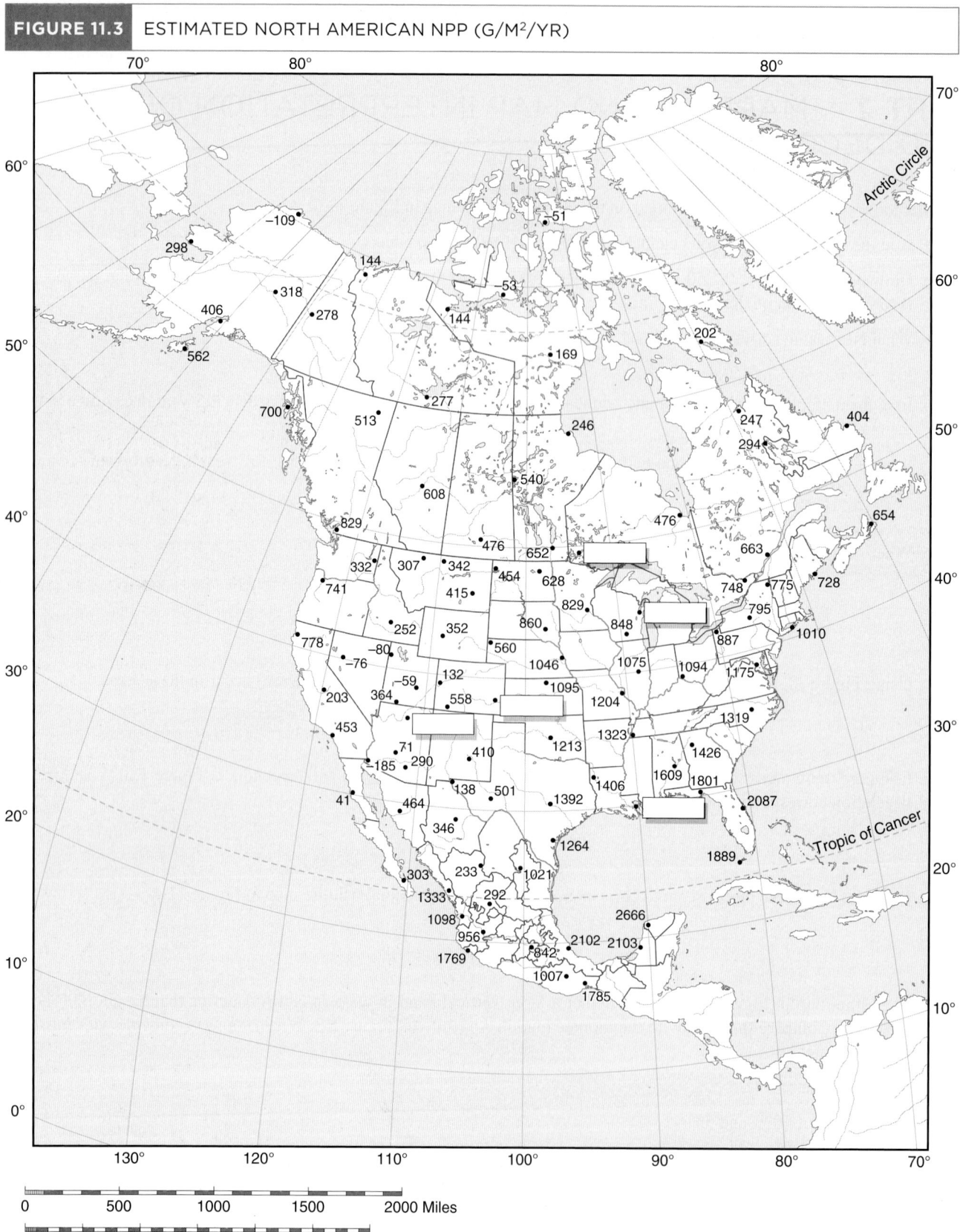

132

5. a. Compare your NPP map in Figure 11.3 with the vegetation maps in Appendix E, Figure E.3 and Appendix F, Figure F.3. What would be good approximate cutoff values for:

	Annual NPP (g/m²/yr)	Decomposition Rate (%/yr)
Tundra	_____	_____
Deciduous forest	_____	_____
Desert	_____	_____

b. Tundra and desert latitudes are quite different, but their NPP and decomposition rates are similar. What climatic conditions account for this similarity?

6. a. The NPP isolines in some areas on your Figure 11.3 map should pack quite closely together, indicating a sharp change of growing conditions within a short distance. For example, notice the drastic change of the NPP index between La Paz and Palos Blancos, Mexico (see Figure 11.5 for their locations), over a distance comparable with Washington, D. C. to New York City. Two other areas with high production gradients are listed below (see Figure 11.5 for their locations). Determine their separating distances, NPP *differences*, and production gradients (equation 11.2) to indicate how sharply growing conditions change. Use the physiographic map in Figure 11.5 to identify landscape features that coincide with these steep NPP gradients. Enter this information in Table 11.2.

TABLE 11.2 LOCATIONS OF STEEP NPP GRADIENTS

From	To	Distance (km)	NPP Difference (g/m²/yr)	Production Gradient (ΔNPP/km)	Physical Feature
La Paz	Palos Blancos	500	1030		salt-water gulf
Eureka	Wendover				
Corpus Christi	El Paso				

b. Why should these types of features exhibit steep production gradients?

7. a. Consult your maps, Table 11.4, and use your knowledge of PE and AE to interpret the data in Table 11.3. Pick a likely North American place (town, state, region, etc.) where you might expect each of the following sets of annual conditions to occur.

TABLE 11.3 EXPECTED NORTH AMERICAN LOCATIONS

Approximate PE (mm)	Approximate AE (mm)	Approximate NPP (g/m²/yr)	Decay Time (yr)	Vegetation	Possible North American Place
550	400	500	2.4	grassland	eastern Colorado, western Nebraska
775	775	1400	1.5	mixed forest	
1275	180	125	3.8	desert	
250	250	200	3.1	tundra	
550	500	775	2.0	evergreen forest	
650	650	1125	1.7	deciduous forest	

b. What sequence of vegetation should you expect to cross if you travel straight from Corpus Christi to El Paso (see the biomes map in Appendix E, Figure E.3)? What probably accounts for this vegetation transition?

c. What sequence of vegetation should you expect to cross if you travel straight from Eureka to Wendover (see the biomes map in Appendix F, Figure F.3)? What probably accounts for this vegetation transition?

FIGURE 11.4 NORTH AMERICA ENVIRONMENTAL CHARACTERISTICS

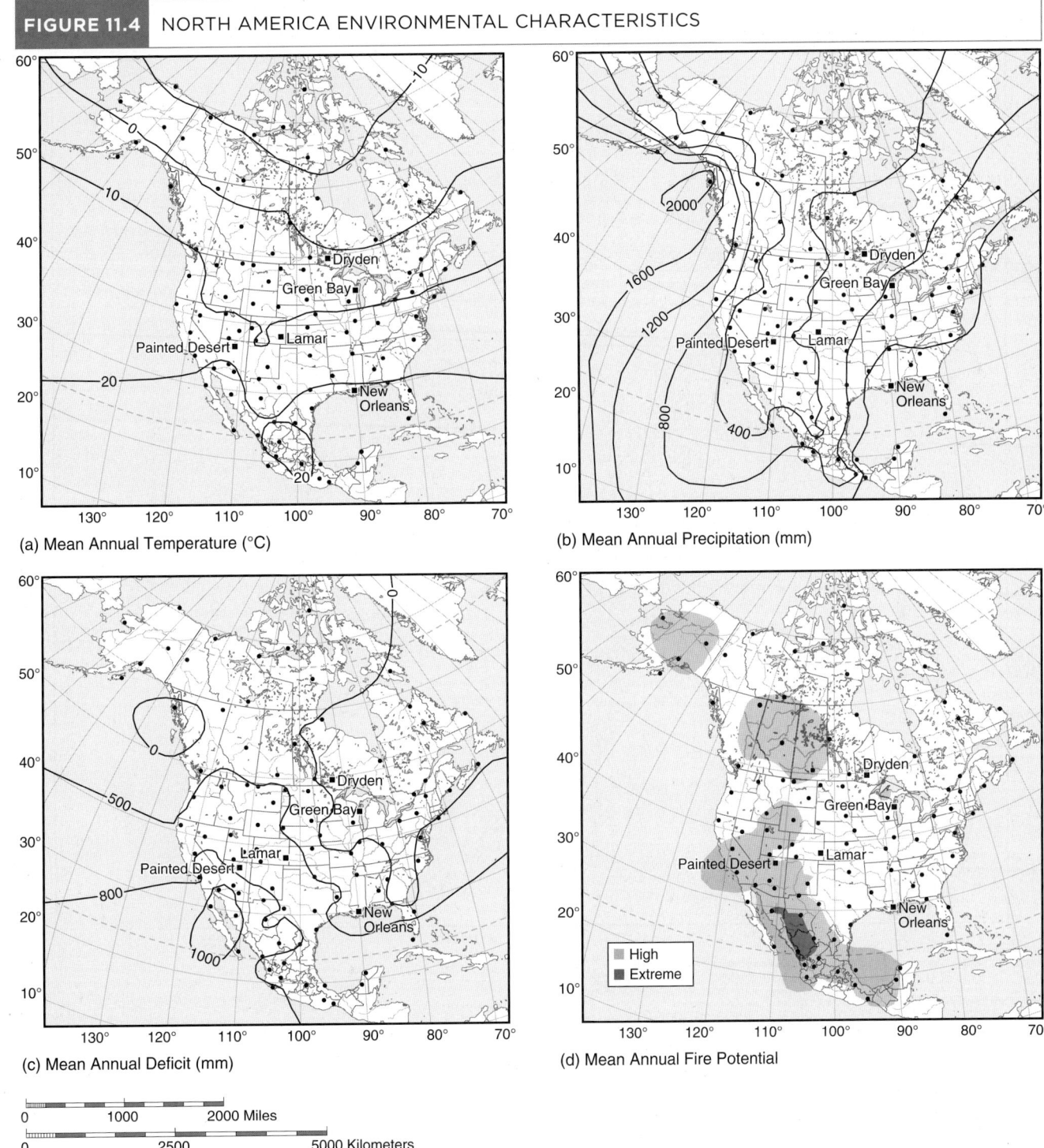

(a) Mean Annual Temperature (°C)
(b) Mean Annual Precipitation (mm)
(c) Mean Annual Deficit (mm)
(d) Mean Annual Fire Potential

135

FIGURE 11.5 NORTH AMERICA PHYSIOGRAPHY

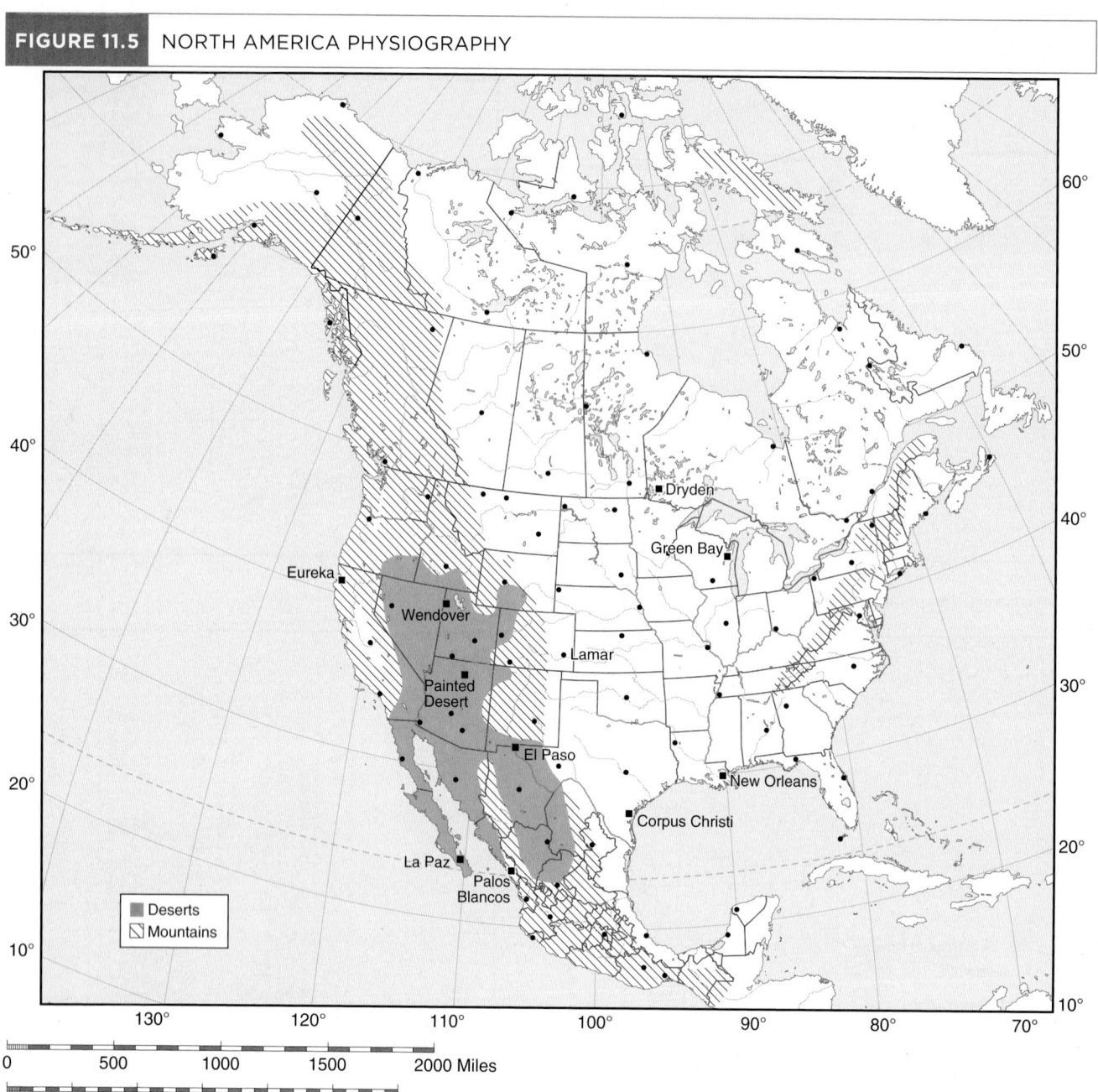

TABLE 11.4 — NORTH AMERICAN AE, NPP, AND DECOMPOSITION RATES IN FINE SAND SOIL

#	Site	Location	Annual AE (mm)	Annual NPP (g/m^2/yr)	Decomposition Rate (%/yr)
1	Nome, Alaska	65°N,165°W	290	298	35
2	Point Barrow, Alaska	71°N,156°W	113	-109	22
3	Fairbanks, Alaska	65°N,148°W	299	318	36
4	Anchorage, Alaska	61°N,150°W	337	406	39
5	Kodiak, Alaska	58°N,153°W	405	562	43
6	Sitka, Alaska	57°N,135°W	466	700	47
7	Dawson, Yukon	64°N,139°W	282	278	35
8	Inuvik, Northwest Territory	69°N,134°W	223	144	30
9	Kugluktuk, Nunavut	68°N,115°W	223	144	30
10	Quamani 'tuaq, Nunavut	64°N,96°W	234	169	31
11	Fort Nelson, British Columbia	59°N,123°W	384	513	42
12	Vancouver, British Columbia	49°N,123°W	522	829	51
13	Hay River, Northwest Territory	60°N,117°W	281	277	35
14	Edmonton, Alberta	54°N,113°W	426	608	45
15	Moose Jaw, Saskatchewan	50°N,106°W	368	476	41
16	Churchill, Manitoba	59°N,94°W	268	246	34
17	Winnipeg, Manitoba	50°N,97°W	445	652	46
18	Moosonee, Ontario	51°N,81°W	368	476	41
19	Ottawa, Ontario	45°N,76°W	487	748	48
20	Kuujjuaq, Quebec	58°N,68°W	268	247	34
21	Schefferville, Quebec	55°N,67°W	289	294	35
22	Quebec City, Quebec	47°N,71°W	450	663	46
23	Cartwright, Newfoundland	54°N,57°W	336	404	38
24	Portland, Oregon	46°N,123°W	484	741	48
25	Fresno, California	37°N,120°W	249	203	32

(continued)

| TABLE 11.4 | NORTH AMERICAN AE, NPP, AND DECOMPOSITION RATES IN FINE SAND SOIL *(CONTINUED)* |

#	Site	Location	Annual AE (mm)	Annual NPP (g/m²/yr)	Decomposition Rate (%/yr)
26	Los Angeles, California	34°N,118°W	358	453	40
27	Lovelock, Nevada	40°N,119°W	127	-76	23
28	Hailey, Idaho	44°N,114°W	270	252	34
29	Spokane, Washington	48°N,118°W	305	332	36
30	Cut Bank, Montana	49°N,112°W	294	307	36
31	Harve, Montana	49°N,110°W	310	342	37
32	Miles City, Montana	46°N,106°W	342	417	39
33	Harrison, Nebraska	43°N,104°W	405	560	43
34	Lander, Wyoming	43°N,108°W	314	352	37
35	Wendover, Utah	41°N,114°W	125	-80	23
36	Hanksville, Utah	38°N,111°W	135	-59	23
37	Kanab, Utah	37°N,113°W	319	364	37
38	Grand Junction, Colorado	39°N,109°W	218	132	30
39	Durango, Colorado	37°N,108°W	404	558	43
40	Roswell, New Mexico	33°N,105°W	339	410	39
41	Phoenix, Arizona	34°N,112°W	191	71	28
42	Yuma, Arizona	33°N,115°W	79	-185	19
43	Tucson, Arizona	32°N,111°W	287	290	35
44	El Paso, Texas	32°N,106°W	220	138	30
45	Fort Stockton, Texas	31°N,103°W	379	501	41
46	Corpus Christi, Texas	28°N,97°W	712	1264	62
47	Oklahoma City, Oklahoma	35°N,98°W	690	1213	60
48	Concordia, Kansas	40°N,98°W	638	1095	57
49	Omaha, Nebraska	41°N,96°W	617	1046	56

(continued)

TABLE 11.4   NORTH AMERICAN AE, NPP, AND DECOMPOSITION RATES IN FINE SAND SOIL *(CONTINUED)*

#	Site	Location	Annual AE (mm)	Annual NPP (g/m²/yr)	Decomposition Rate (%/yr)
50	Williston, North Dakota	48°N,104°W	358	454	40
51	Devil's Lake, North Dakota	48°N,99°W	434	628	45
52	Minneapolis, Minnesota	45°N,93°W	522	829	51
53	Madison, Wisconsin	43°N,89°W	530	848	51
54	St. Louis, Missouri	39°N,90°W	686	1204	60
55	Memphis, Tennessee	35°N,90°W	738	1323	63
56	Key West, Florida	25°N,82°W	985	1889	76
57	Tallahassee, Florida	30°N,84°W	946	1801	74
58	Atlanta, Georgia	34°N,84°W	783	1426	65
59	Washington, D. C.	39°N,77°W	673	1175	59
60	Cincinnati, Ohio	39°N,85°W	638	1094	57
61	Erie, Pennsylvania	42°N,80°W	547	887	52
62	Cortland, New York	43°N,76°W	507	795	50
63	Riverhead, New York	41°N,73°W	601	1010	55
64	Plattsburgh, New York	45°N,73°W	498	775	49
65	Arcadia Nat. Park, Maine	44°N,68°W	478	728	48
66	La Paz, Baja California Sur	24°N,110°W	293	303	35
67	Tepic, Nayarit	22°N,105°W	640	1098	58
68	Mexico City, Mexico	19°N,99°W	528	842	51
69	Monterrey, Neuvo Leon	26°N,100°W	606	1021	56
70	Merida, Yucatan	21°N,90°W	1324	2666	91
71	Champoton, Campeche	19°N,91°W	1079	2103	80
72	Decatur, Illinois	40°N,89°W	630	1075	57
73	El Oregano, Sonora	29°N,111°W	363	464	40

(continued)

TABLE 11.4 NORTH AMERICAN AE, NPP, AND DECOMPOSITION RATES IN FINE SAND SOIL (CONTINUED)

#	Site	Location	Annual AE (mm)	Annual NPP (g/m²/yr)	Decomposition Rate (%/yr)
74	Guadalajara, Jalisco	21°N,103°W	576	956	54
75	Palos Blancos, Sinaloa	25°N,107°W	742	1333	63
76	Zacatecas, Zacatecas	23°N,103°W	288	292	35
77	Oaxaca, Oaxaca	17°N,97°W	600	1007	55
78	Veracruz, Veracruz	19°N,96°W	1078	2102	80
79	Matias Romero, Oaxaca	17°N,95°W	940	1785	73
80	Manzanillo, Colima	19°N,104°W	933	1769	73
81	Montgomery, Alabama	32°N,86°W	863	1609	70
82	Shreveport, Louisiana	33°N,94°W	774	1406	65
83	Cambridge Bay, Nunavut	69°N,105°W	137	-53	24
84	Iqaluit, Nunavut	64°N,69°W	249	202	32
85	Resolute, Nunavut	75°N,95°W	138	-51	24
86	Flin Flon, Manitoba	55°N,102°W	396	540	43
87	Sydney, Nova Scotia	46°N,60°W	446	654	46
88	Eureka, California	41°N,124°W	500	778	49
89	Raleigh, North Carolina	36°N,79°W	736	1319	63
90	Titusville, Florida	29°N,81°W	1072	2087	80
91	Austin, Texas	30°N,98°W	768	1392	65
92	Mitchell, South Dakota	44°N,98°W	536	860	51
93	El Rosario, Baja California	30°N,116°W	178	41	27
94	Cuidad Lerdo, Durango	24°N,104°W	262	233	33
95	Chihuahua, Chihuahua	29°N,106°W	311	346	37

EXERCISE 12 · VEGETATION FORM AND RANGE

PURPOSE

The purpose of this exercise is to examine global patterns of the five terrestrial biomes by isolating single attributes from a multivariate classification scheme.

LEARNING OBJECTIVES

By the end of this exercise you should be able to

- describe vegetation characteristics of the five terrestrial biomes, using Küchler's classification;
- compare and contrast vegetation characteristics associated with particular biomes;
- describe the spatial distributions of particular biomes; and
- predict likely locations for dispersal of plants having particular combinations of vegetation characteristics.

INTRODUCTION

After three decades of research, August Wilhelm Küchler in 1964 published a map of United States vegetation that employed a complex new classification system. Today, with subsequent refinement, Küchler's scheme remains a standard means of depicting the spatial distribution of dominant vegetation characteristics. This exercise separates Küchler's single complex code into the four vegetation characteristics of form, stature, habit, and density for the purpose of examining vegetation ranges.

A **range** is the geographic area that a particular kind of organism occupies. Ranges change as organisms move into or out of an area. Such dispersal into new ranges sometimes occurs rapidly, or it may be a very gradual process. Although related, the terms *dispersion* and *dispersal* do not have quite the same meaning. **Dispersal** is the process by which organisms permanently relocate into new areas. This may include dispersal of seeds by traveling animals, water currents, or the wind. Dispersion is the resultant pattern of these organisms' spatial distribution; that is, it is a characteristic arrangement for multiple individuals within the range of an organism. There is no scale limitation to dispersion; it can mean the linear clustering of a water-loving species along the banks of a single stream, or it can focus upon global patterns of disjunct (separate) areas occupied by the same kind of organism.

Vegetation Characteristics

The four vegetation characteristics (physiognomy) are plant form, stature, habit, and density.

There are many ways to describe plant **form**, which means the physical characteristics of plants. For foliage, Küchler differentiated between *broadleaf, needleleaf, mixed broadleaf and needleleaf, grass,* and *lichen and herb* forms. *Barren* indicates an absence of vegetation.

Stature refers to the typical height of a mature plant. Küchler distinguished between *trees* (woody single-stemmed plants exceeding 1.5 meters in height), *shrubs* (woody multiple-stemmed plants) that are either taller or shorter than one meter, and *grass or short herbs.* Again, *barren* indicates an absence of vegetation.

Küchler further distinguished dominant plants by their **habit,** or characteristic seasonal growth action. The classes are *evergreen* (no season of foliage shedding), *deciduous* (regular seasonal foliage shedding), *semideciduous* (irregular timing of foliage loss when adverse conditions occur), *mixed habits,* and perennial *subsurface rootstock or seed habits.*

Finally, Küchler classified vegetation spatially by **density,** which is the characteristic spacing and dispersion of individual plants. *Continuous* means unbroken foliage cover, *patches* are small clusters separated by other ground cover, *isolated* is wide proximity that precludes contact between individual neighbors, and *barren* indicates a lack of apparent vegetation.

The Five Terrestrial Biomes

Organisms with similar characteristics often occupy similar ranges, and this is the basis of Küchler's classification. Küchler focused primarily on the physiognomy (form, stature, habit, density) of plants, and these often are adaptive responses to habitat conditions, so it becomes possible to correlate his vegetative regions with conventional biomes. A **biome** is a characteristic assemblage of ecosystems, including the nonliving components such as climate and terrain. There are five major land biomes: forest, savanna, grassland, desert, and tundra.

Forest, by the definition of the United States Geological Survey, means a continuous canopy of trees. The forest biome covers approximately one-third of the world's land areas. Some forests have few species, but many others are noteworthy for their extreme **diversity** (species richness).

Most forests (and also the other biomes), in fact, are mosaics of multiple interactive ecosystems. Forests tend to be more diverse at low latitudes (0–30°), in regions of large topographic relief, or when they are of greater age. A hectare (2.47 acres) of tropical rainforest often holds over 150 tree species but might have only a single individual of each. Midlatitude (30–60°) forests tend to have moderate diversity, and high latitude (60–80°) or high altitude (over 3000 meter elevation) forests characteristically exhibit rather low diversity.

Savanna (sometimes labeled tropical grassland) covers about one-sixth of the earth's land area. "Savanna" is a Native American word meaning "grassy area not covered with forest," but modern usage restricts the term to the extensive tropical and subtropical areas having one of three formation groups: (1) tall grass with isolated trees (*parkland*), (2) grass with thorny shrubs and low trees (*woodland*), or (3) discontinuous short grass with occasional trees (*scrub savanna*). All forms of savanna occur in the areas having a tropical wet and dry climate that exist poleward of the tropical rainforests and usually equatorward of subtropical deserts. The specific formation group that occurs at a place is mainly a response to the duration of the dry season; regions with longer dry seasons tend to produce shorter, patchier vegetation cover.

Grassland in biogeography usually refers to midlatitude interior regions where effective moisture is too low to support forest but not so low as to have desert, even though grasses grow in all other communities and, in fact, dominate as ground cover in some of them. Grassland includes the formation groups of continuous tall-grass *prairie* (typically 1 to 1.5 meters high) in somewhat moister areas, and discontinuous short-grass *steppe* (<0.2 meters) in the more arid parts of grassland regions. Transitional between the two subtypes is medium (or mixed) grassland. Short-grass steppe may superficially resemble scrub savanna, but steppe lacks the isolated trees found in scrub savanna.

Desert is a large biome, comprising about one-fourth of all land, and occurring on all continents except Antarctica. Precipitation is not only low but also erratic; the variability of precipitation from season to season or year to year is high. Due to the dry clear air, temperatures in most deserts exhibit large daily extremes, and at higher latitudes the annual variations of temperature may also be very large. Vegetation in deserts is discontinuous (much bare soil shows between individual plants), and annual organic productivity is extremely low—in some barren deserts, it is virtually nil. Deserts occur in four typical situations: beneath subtropical highs, in rainshadows, in the deep interior of continents, or along coasts that parallel cold ocean currents where prevailing surface winds usually blow offshore (from land, toward the sea).

Tundra occurs in the absolutely treeless biome of high latitude (arctic) or high altitude (alpine) regions. The common characteristic is low temperature. Although moisture presence can vary from hyperarid to extremely wet, most water remains frozen most of the time and therefore is unusable to most plants. Only shallow-rooted vegetation can obtain liquid moisture during the brief thaw season, and often for only a few weeks of the entire year. The freeze-thaw cycles also lead to unstable heaving soils, and slope instability further limits rooting effectiveness in alpine tundra. Arctic tundra has very long, or 24-hour, exposure to low-intensity sunlight during summer, but its continuously low temperatures create permafrost (permanently subfreezing soil temperature) below the thaw layer, and ground ice if water is present in the permafrost.

IMPORTANT TERMS, PHRASES, AND CONCEPTS

range	biome
form	forest
stature	savanna
habit	grassland
density	desert
dispersal	tundra
diversity	

REFERENCE

Küchler, August Wilhelm. 1964. *Potential Natural Vegetation in the Conterminous United States* (map). New York: American Geographical Society, Special Publication No. 36.

Name: _____   Section: _____

PART 1 • RANGE AND FORM

1. Use Appendix E, Figures E.1 through E.4 and Appendix F, Figure F.3 to classify the four vegetation characteristics for each of the stations in Table 12.1, using the codes below. Cincinnati is complete as an example.

TABLE 12.1 VEGETATION PHYSIOGNOMY OF SELECTED LOCATIONS

Station	Location	Form	Stature	Habit	Density	Biome
Cincinnati, Ohio	39°N, 85°W	B	t	D	c	forest
Irkutsk, Russia	52°N, 104°E					
Manaus, Brazil	3°S, 60°W					
Kolkata, India	23°N, 88°E					
Cape Town, South Africa	34°S, 18°E					
Merida, Yucatan	21°N, 90°W					
Brasilia, Brazil	16°S, 48°W					
Mombasa, Kenya	4°S, 40°E					
Darwin, Australia	12°S, 131°E					
Mar del Plata, Argentina	38°S, 58°W					
Moose Jaw, Saskatchewan	50°N, 106°W					
Odesa, Ukraine	46°N, 31°E					
Alice Springs, Australia	24°S, 134°E					
Zaltan, Libya	28°N, 20°E					
Torreón, Mexico	26°N, 103°W					
Barrow, Alaska	71°N, 156°W					
Qamani'tuaq, Nunavut	64°N, 96°W					
Anadyr', Russia	65°N, 178°E					

Dominant Form

B = Broadleaf

N = Needleleaf

M = Mixed Broadleaf and Needleleaf

G = Grass

L = Lichens and nongrass Herbs

b = barren (nonvegetated)

Dominant Habit

E = Evergreen

D = Deciduous

S = Semideciduous

R = Rootstock or Seeds (subsurface)

b = barren (nonvegetated)

Maximum Stature

t = trees > 1.5 m height

s = shrub >1 m height

z = shrub <1 m height

g = grass or low herbs

b = barren (nonvegetated)

Prevailing Density

c = continuous

p = patches

i = isolated

b = barren (nonvegetated)

2. Using the information in Table 12.1, specify the most common physiognomy of the five terrestrial biomes.

	Form	Stature	Habit	Density
Forest	_____	_____	_____	_____
Savanna	_____	_____	_____	_____
Grassland	_____	_____	_____	_____
Tundra	_____	_____	_____	_____
Desert	_____	_____	_____	_____

3. In question 2, you listed the primary physiognomy of the five biomes. What characteristic(s) seems most common at the following locations? Refer to Appendix E, Figures E.1 through E.4 and Appendix F, Figure F.3.

	Under 30° latitude	Over 30° latitude	Northern Hemisphere	Southern Hemisphere
Form	_____	_____	_____	_____
Habit	_____	_____	_____	_____
Stature	_____	_____	_____	_____
Density	_____	_____	_____	_____

4. According to Table 12.1 and the maps in Appendix E, Figures E.1 through E.4, and Appendix F, Figure F.3:

   a. What one characteristic (form, stature, habit, or density) do all forests appear to share? Be specific.

   b. What biome coincides with areas of sparse vegetation (patchy or isolated low shrubs), or lacks surface vegetation entirely?

   c. What one characteristic (form, stature, habit, or density) of forests usually differs from savanna, grassland, desert, and tundra? Be specific.

d. What characteristic(s) distinguishes savanna from grassland? Be specific.

  e. What characteristic(s) distinguishes grassland from desert? Be specific.

  f. What characteristic(s) distinguishes desert from tundra? Be specific.

5. In a general fashion, describe the locations where the forest biome occurs (use Appendix F, Figure F.3).

6. In a general fashion, describe several locations where deciduous forest occurs (use Appendix E, Figures E.2 and E.3 and Appendix F, Figure F.3).

7. a. Where is there a large area of needleleaf deciduous forest (use Appendix E, Figures E.1 and E.3)?

   b. Speculate as to why being both needleleaf and deciduous might be useful for surviving in this area.

Name: _____   Section: _____

PART 2 ✦ RANGE AND DISPERSAL

1. Sagebrush and its taxonomic relatives (*Artemisia* spp.) are widespread broadleaf shrubs that tend to occur in patches or as isolated individuals and are frost-tolerant. Using Appendix E, Figures E.1 through E.4 and Appendix F, Figure F.3, hypothesize three areas (e.g., Japan? Cuba?) where sagebrush could exist.

2. What biome(s) (Appendix F, Figure F.3) most corresponds with the three areas that you selected?

3. Figure 12.1 shows the actual distribution of sagebrush. Does this pattern confirm or refute your hypothesis in question 1?

4. a. What latitude zones (low = 0° to 30°, middle = 30° to 60°, high >60°) does sagebrush occupy?

   b. Does sagebrush have more of a coastal or a continental interior distribution?

5. a. Describe the climatic conditions that should be favorable for sagebrush to dominate.

   b. On the basis of its actual distribution (Figure 12.1), what climate conditions seem to coincide with an absence of sagebrush?

   c. What biomes (Appendix F, Figure F.3) seem to coincide with an absence of sagebrush?

6. Considering the previous questions, which of the following range descriptors seem applicable to sagebrush? (Check all that apply.)

✓	Range Type	Description
	Cosmopolitan	occurs almost anywhere, with very few limitations
	Pantropical	occurs around the world, within the tropical latitudes
	Pantemperate	occurs around the world, within the middle latitudes
	Panarctic	occurs around the world, within the northern high latitudes
	Panantarctic	occurs around the world, but limited to the southern high latitudes
	Amphitropical	occurs above the tropical latitudes, in northern and southern hemispheres

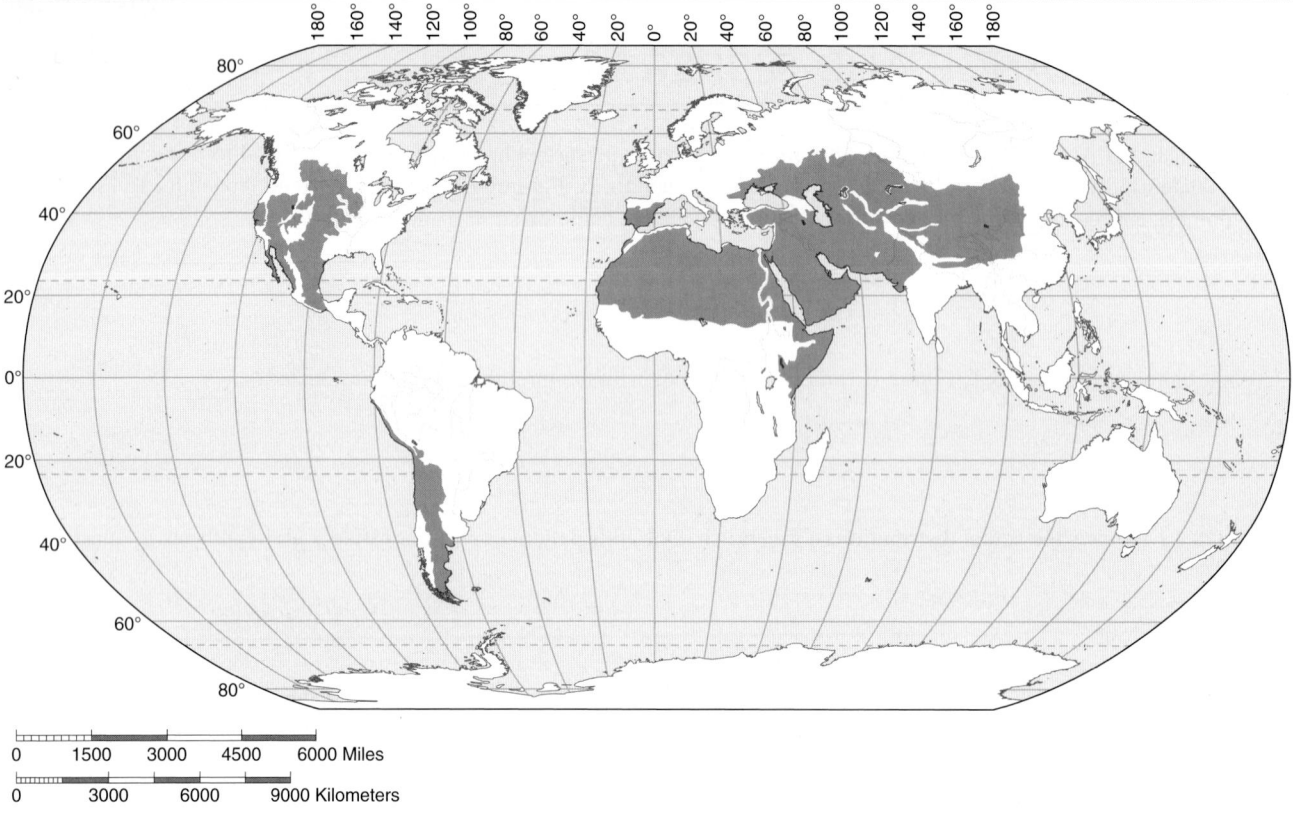

FIGURE 12.1 CONTEMPORARY SAGEBRUSH (*ARTEMISIA* SPP.) RANGE (SHADED)

7. Notice that sagebrush occurs in southwestern South America, where it is more recent and far from its ranges on other continents. How do you suppose it got to South America?

8. Where else might sagebrush be able to establish, if it relocated to a nonnative region with suitable habitat? Refer to Appendix E, Figures E.1 through E.4 and Appendix F, Figure F.3.

EXERCISE 13 ♦ BIOCLIMATIC TRANSECTS

PURPOSE

The purpose of this exercise is to examine systematic changes in environmental conditions as latitude or altitude change, and to consider the degree of equivalency in the effect of latitude and altitude upon environmental variability.

LEARNING OBJECTIVES

By the end of this exercise you should be able to

- use Holdridge's Life Zone triangle to predict vegetation communities at a site;
- describe latitudinal changes of life zones along the North American west coast and within midcontinental North America;
- describe altitudinal changes of life zones ascending western North American mountain ranges;
- compare and contrast latitudinal changes in temperature and moisture in the midcontinent with the west coast of North America; and
- explain how latitudinal and altitudinal sequences of life zones correspond with latitudinal and altitudinal changes in temperature and moisture.

INTRODUCTION

Vegetation differs from place to place on earth, and any single species of plant usually has its distribution restricted to an area, or geographic range, where environmental conditions most suit it. In 1947, L. R. Holdridge hypothesized that if plants selectively evolve, then each kind of plant should survive only within its rather narrow climatic limits but typically in conjunction with other plants having similar climatic tolerances, creating natural **communities**. These communities are classifiable by the prevalent plant forms (physical attributes such as shape or size) and habits (rate and seasonality of growth, and longevity). Thus each community reflects the climate within its members' geographical ranges.

Holdridge's scheme was to classify any particular place by those climatic qualities he regarded as being most crucial to plant growth. His goal was to predict extant plant communities based on climatic characteristics, with subsequent verification by site visits. Holdridge selected climatic variables that he believed regulated plant growth and were measurable or calculable from normal climate data. Ultimately, he opted to rely on only three predictor variables: mean annual precipitation (P) to represent moisture availability, mean annual biotemperature (T_B) to represent the heat available during the growing season, and mean annual potential evapotranspiration (PE) to incorporate a hypothetical maximum availability of both moisture and heat.

Holdridge's Climatic Variables

Precipitation (P) is the average annual precipitation, expressed in water equivalency units as depths. P includes the winter months' precipitation on the logic that snowmelt eventually supplies soil water to plants during the growing season.

Biotemperature (T_B) is a calculated value designed to represent the heat conditions only while plants are growing. Since most plants cannot use frozen water for growth, only the mean monthly temperatures (MMT) above fresh water's freezing point (0°C) are used. These mean monthly values are added and then divided by 12. The resulting number is the mean annual biotemperature (T_B):

$$T_B = \frac{\sum MMT > 0}{12} \qquad (13.1)$$

where: T_B = mean annual biotemperature

$\sum$ = sum of the variable that follows

MMT = mean monthly temperature.

Potential evapotranspiration (PE) is the maximum amount of water that a site could yield given its temperature if unlimited water were available at the ground surface for evapotranspiration. Like precipitation (P), this variable is expressed as water equivalency depths. Through many direct trial-and-error observations, Holdridge found a consistent relationship whereby annual PE is 58.43 mm for each degree of biotemperature (T_B). Thus, instead of determining annual PE from a water budget, it may be estimated by the equation:

$$PE = 58.43(T_B) \qquad (13.2)$$

where: PE = annual potential evapotranspiration

T_B = mean annual biotemperature.

Holdridge's ratio (PE/P) is an index relating the amount of moisture that a site could lose (PE) to the amount received as precipitation (P):

$$\text{Holdridge's ratio} = \frac{PE}{P} \qquad (13.3)$$

A ratio value greater than 1.00 means that there is greater loss than receipt of water at the site; the climate is dry. A value less than 1.00 means that the site receives more water than it loses; a wet climate. How dry or wet a site is becomes proportional to the difference between the calculated ratio value and 1.00; for example, a ratio of 2.05 is much drier than a ratio of 1.12.

Predicting Vegetation Communities

The Life Zone Triangle (Figure 13.1) uses the climatic variables to predict vegetation communities, or **life zones.** Within the Life Zone Triangle, the diagonal *dotted* line where Holdridge's ratio equals 1.0 is the boundary between wet (PE < P) and dry (PE > P) climates.

Predict a site's life zone by first locating its mean annual precipitation values outside the left and bottom edges of the triangle; draw a straight diagonal line between these positions. For example, data from Kugluktuk, Nunavut (Table 13.1), reveals a mean annual precipitation of 238 mm. Figure 13.1

| FIGURE 13.1 | THE LIFE ZONE TRIANGLE |

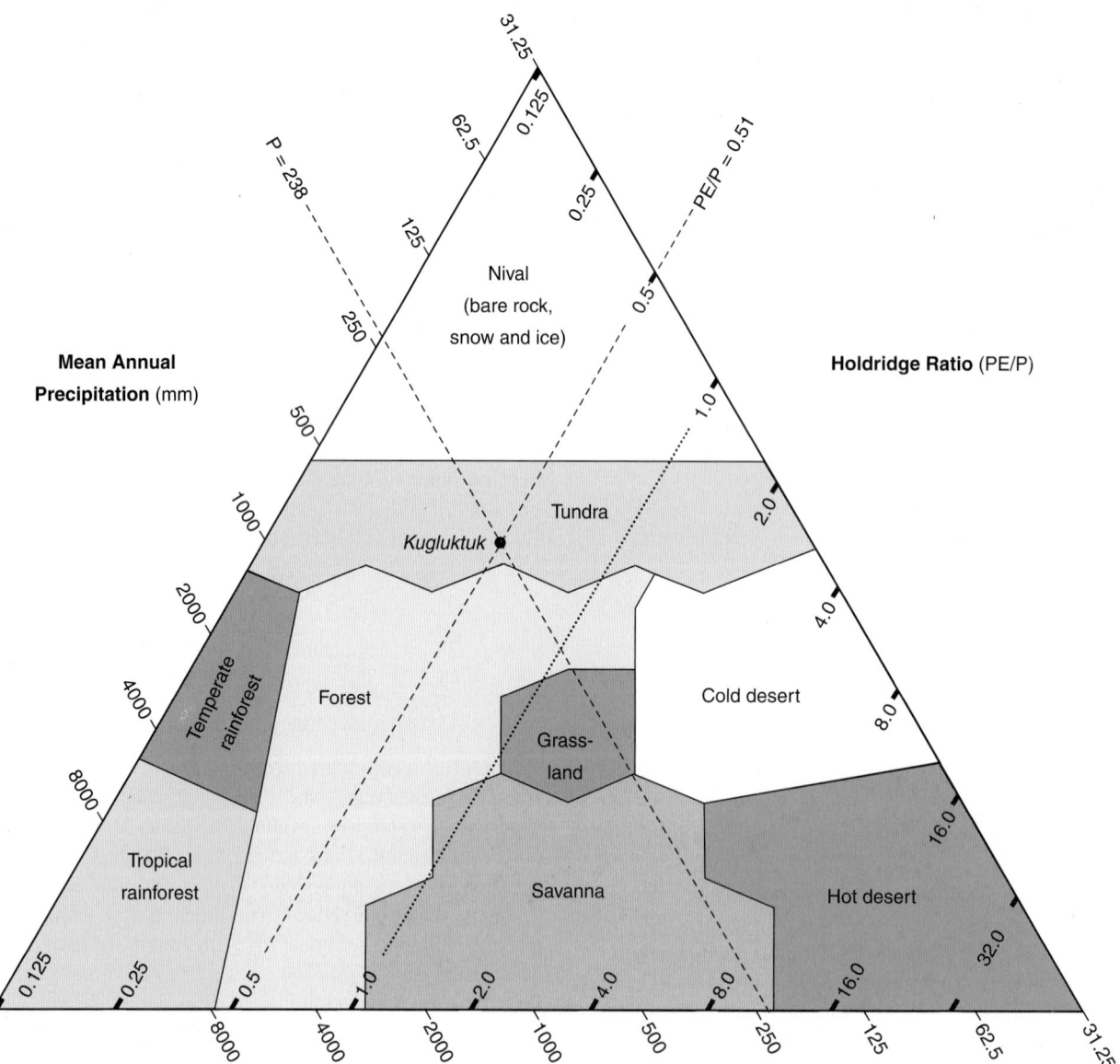

Source: Adapted from Holdridge, 1947.

contains a dashed line connecting the value of 238 on the outside bottom axis to the value of 238 on the left axis.

Adding Kugluktuk's four monthly above-freezing temperatures (Table 13.1) and then dividing by 12 results in a biotemperature of 2.08°C. Substituting this value into equation 13.2 produces an annual PE of 121.53 mm. Thus, Holdridge's ratio (equation 13.3) equals 0.51, indicating that Kugluktuk has a wet climate. Figure 13.1 contains a second dashed line connecting the value of 0.51 for Holdridge's ratio on the inside bottom axis to the value of 0.51 on the inside right axis of the Life Zone Triangle.

The predicted life zone for a selected site is the graph space within which the two lines cross; tundra is the predicted community for Kugluktuk. Label with the site's name the point where the two lines cross.

Confirming Vegetation Predictions

Holdridge verified his predictions by site visits, which is the best method for confirming predictions; however, alternative documentary evidence, such as written site descriptions, maps, or site imagery, also provide a means of verifying predictions. In this exercise we will "verify" predicted vegetation communities by maps and site photographs. Appendix maps and a photograph (Figure 13.3) confirm that Kugluktuk is indeed tundra.

TABLE 13.1 MEAN MONTHLY CLIMATE DATA FOR KUGLUKTUK, NUNAVUT

	Jan	Feb	Mar	Apr	May	Jun	Jul	Aug	Sep	Oct	Nov	Dec	Year
Kugluktuk, Nunavut (68°N, 115°W, elevation 22 m)														
T (°C)	−29	−30	−26	−17	−6	4	10	8	3	−7	−20	−26	−11.33
P (mm)	12	8	13	14	14	18	32	41	30	28	16	12	238

FIGURE 13.2 TRANSECT LOCATIONS AND DATA COLLECTION POINTS

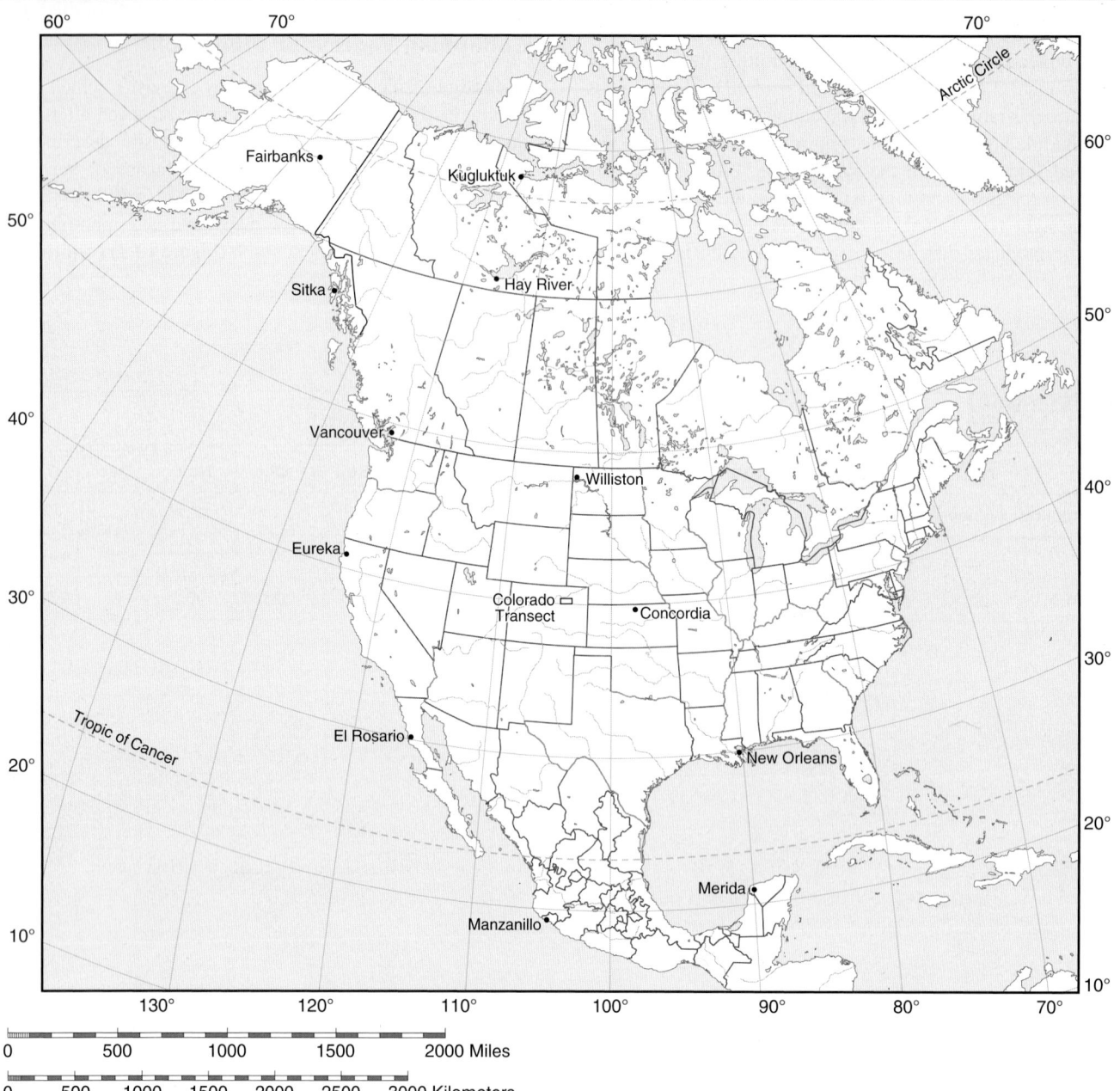

Bioclimatic Transect Data

Figure 13.2 shows two latitudinal transects (a transect is simply a route along which one makes observations). Crossing latitudes are a west-coast transect and a midcontinental transect, each having six observation sites. Both transects extend across almost 10,000 kilometers, but all sites are within 550 meters of sea level.

Figure 13.2 also shows the location of an altitudinal transect in Colorado. Much as climates and vegetation

FIGURE 13.3 TRANSECT SITE PHOTOGRAPHS

Kugluktuk

Hay River

Williston

Concordia

New Orleans

Merida

(a) Midcontinent latitudinal transect

Niwot Ridge

Como Creek

Sugarloaf

Ponderosa

Boulder

Longmont

(b) Altitudinal transect

Sources: Part (a): top right Courtesy of the National Park Service; top middle Courtesy NASA/JPL-Caltech; top left © Norbert Rosing/Getty; bottom right © Kristi J. Black/Corbis; bottom middle Courtesy of Gulf Restoration Network/www.healthygulf.org; bottom left ©Courtesy of David Chapman/www.stormdude.com; (b): © Neil Heywood.

differ across latitude, so bioclimatic communities at unlike altitudes above sea level also vary. A good illustration of such altitudinal (sometimes called "vertical") zonation occurs on the eastern slope of the Colorado Front Range. At this continental interior location, the high plains steppe environment abruptly ends and there is a rapid transition through several distinctive bioclimatic zones as one moves up into the Rocky Mountains. The altitudinal transect here consists of six sites along an east-west line that is just 43 kilometers (27 miles) long, but entails a vertical rise of 2234 meters (7330 feet) from the lowest site (Longmont, 1509 meters above sea level) to the highest site (Niwot Ridge, 3743 meters above sea level).

Table 13.2 contains bioclimatic information for the west coast transect sites. Table 13.3 contains bioclimatic characteristics for the six sites along the Colorado altitudinal transect. Table 13.4 contains climate data for the five additional sites along the midcontinental transect, data you shall use to complete a third bioclimatic information summary in Table 13.5.

TABLE 13.2 WEST COAST LATITUDINAL TRANSECT BIOCLIMATIC CHARACTERISTICS

	Fairbanks, Alaska	Sitka, Alaska	Vancouver, British Columbia	Eureka, California	El Rosario, Baja California Norte	Manzanillo, Colima
Latitude	65°N	57°N	49°N	41°N	30°N	19°N
Altitude	147 m	20 m	3 m	13 m	82 m	3 m
MAT (°C)	−3.2°	6.3°	9.8°	11.1°	22.5°	26.5°
P (mm)	317	2300	1112	978	178	1017
T_B (°C)	4.92°	6.33°	9.82°	11.14°	22.46°	26.48°
PE (mm)	287.28	369.57	573.79	651.01	1312.24	1547.42
Holdridge ratio PE/P	0.91	0.16	0.52	0.67	7.36	1.52
Life Zone	forest	temperate rainforest	forest	forest	hot desert	savanna
Map Community	taiga forest	temperate rainforest	temperate rainforest	temperate rainforest	hot desert	tropical forest
Photograph Community	taiga forest	temperate rainforest	temperate rainforest	temperate rainforest	hot desert	tropical forest

TABLE 13.3 COLORADO ALTITUDINAL TRANSECT BIOCLIMATIC CHARACTERISTICS

	Niwot Ridge	Como Creek	Sugarloaf	Ponderosa	Boulder	Longmont
Latitude	40.1°N	40.1°N	40.1°N	40.1°N	40.1°N	40.2°N
Altitude	3743 m	3048 m	2591 m	2195 m	1652 m	1509 m
MAT (°C)	−3.8°	1.3°	5.6°	8.2°	10.2°	9.3°
P (mm)	926	771	582	579	473	327
T_B (°C)	2.25°	3.92°	6.58°	8.33°	10.23°	9.75°
PE (mm)	131.47	229.05	384.97	486.72	597.74	569.69
Holdridge ratio PE/P	0.14	0.30	0.66	0.84	1.26	1.74
Life Zone	Tundra	Forest	Forest	Forest	Grassland	Grassland
Graph Community	Tundra	Forest	Forest	Forest	Grassland	Grassland
Photograph Community	Tundra	Forest	Forest	Forest	Forest	Grassland

TABLE 13.4 MEAN MONTHLY CLIMATE DATA FOR THE MIDCONTINENTAL TRANSECT

	Jan	Feb	Mar	Apr	May	Jun	Jul	Aug	Sep	Oct	Nov	Dec	Year
Hay River, Northwest Territories (60°N, 117°W, elevation 166 m)														
T (°C)	−25	−22	−17	−4	5	11	16	14	8	0	−12	−21	−3.92
P (mm)	19	17	16	16	22	31	40	43	38	21	31	22	326
Williston, North Dakota (48°N, 104°W, elevation 549 m)														
T (°C)	−13	−11	−4	6	12	17	21	20	14	7	−2	−9	4.83
P (mm)	13	11	19	29	47	77	49	37	31	19	14	13	358
Concordia, Kansas (40°N, 98°W, elevation 454 m)														
T (°C)	−3	0	5	12	17	23	26	25	20	14	5	0	12.00
P (mm)	15	22	40	57	105	111	86	82	69	47	26	18	677
New Orleans, Louisiana (30°N, 90°W, elevation 1 m)														
T (°C)	12	14	17	20	24	27	28	28	26	22	16	13	20.58
P (mm)	118	119	129	124	117	145	171	156	132	84	100	122	1518
Merida, Yucatan (21°N, 90°W, elevation 22 m)														
T (°C)	23	24	26	27	28	28	27	27	27	26	24	23	25.83
P (mm)	63	92	24	47	96	189	130	150	223	269	174	130	1586

In addition to the climatic variables used by Holdridge, Tables 13.2 – 13.5 contain another variable, mean annual temperature (MAT):

$$MAT = \sum MMT/12 \qquad (13.4)$$

where: MAT = mean annual temperature

$\sum$ = sum of the variable that follows

MMT = mean monthly temperature.

Figure 13.3 contains photographs of the six midcontinental sites. Figure 13.4 is a vegetation profile of the altitude transect, from the south view. Use all table data and figures to answer the exercise questions.

IMPORTANT TERMS, PHRASES, AND CONCEPTS

community

precipitation (P)

biotemperature (T_B)

potential evapotranspiration (PE)

Holdridge's ratio (PE/P)

life zone

REFERENCE

Holdridge, L. R. 1947. Determination of world plant formations from simple climatic data. *Science* 105: 367–368.

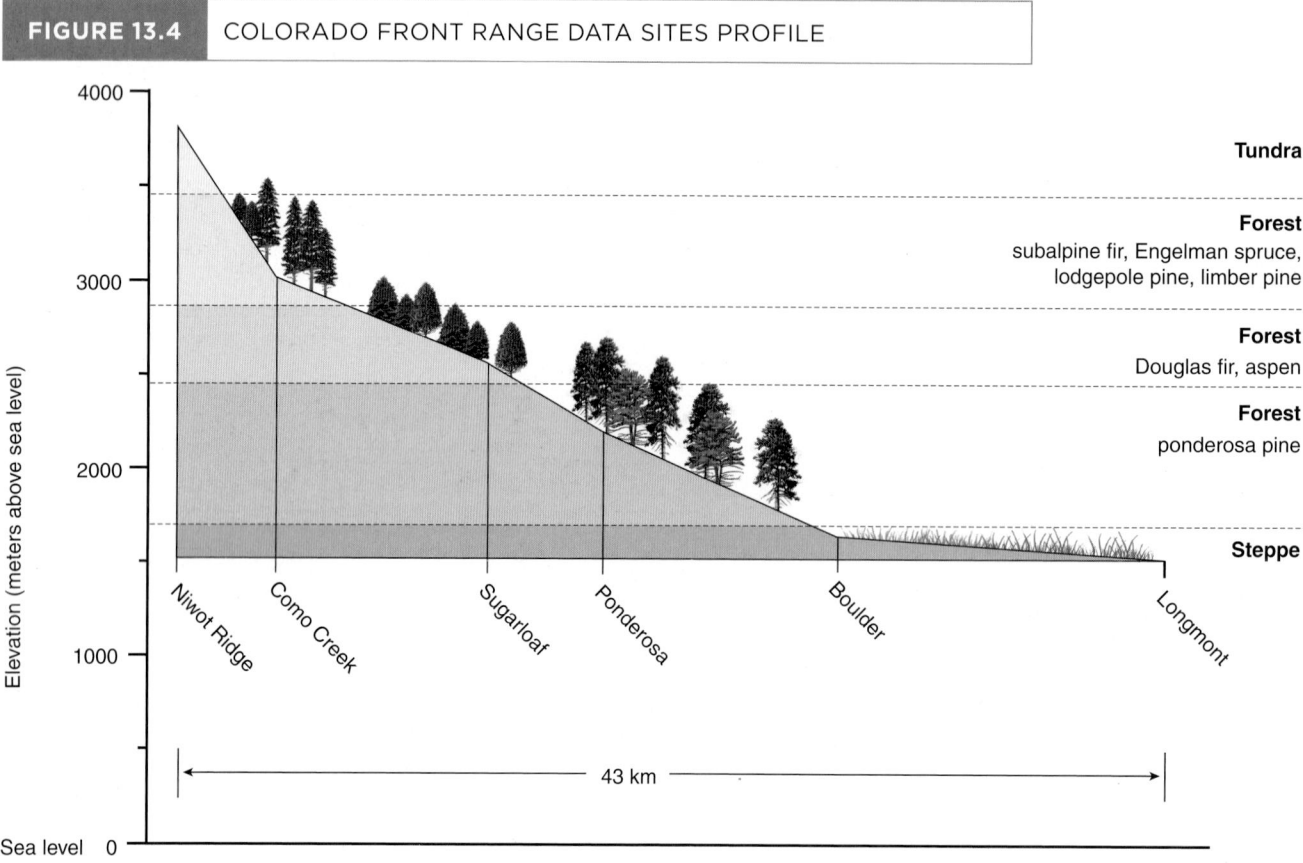

FIGURE 13.4 COLORADO FRONT RANGE DATA SITES PROFILE

Name: _____   Section: _____

BIOCLIMATIC TRANSECTS

Use data from Tables 13.2, 13.3, and 13.4, as well as all the figures in the Introduction, to answer the following questions.

1. Complete Table 13.5 for sites along the midcontinent transect using climatic data from Table 13.4. Round all climatic index variables to the nearest .01. Then use the Life Zone Triangle (Figure 13.1) to predict the vegetation community for each site. Use the vegetation maps (Appendix E, Figure E.3 and Appendix F, Figure F.3) and the photographs (Figure 13.3) to confirm or refute the Life Zone Triangle predictions. Kugluktuk entries are complete as an example.

TABLE 13.5 MIDCONTINENTAL LATITUDINAL TRANSECT BIOCLIMATIC CHARACTERISTICS

	Kugluktuk, Nunavut	Hay River, Northwest Territories	Williston, North Dakota	Concordia, Kansas	New Orleans, Louisiana	Merida, Yucatan
Latitude	68°N					
Altitude	22 m					
MAT (°C)	−11.33°					
P (mm)	238					
T_B (°C)	2.08°					
PE (mm)	121.53					
Holdridge ratio PE/P	0.51					
Life Zone	tundra					
Map Community	tundra					
Photograph Community	tundra					

2. a. How many of the site life zone predictions were correct?

   West Coast _____   Midcontinent _____

   b. List any erroneous sites:

   West Coast

   Midcontinent

   c. What might account for the mispredictions at any erroneous sites?

157

3. What generalizations can you make about the progression of life zones (Figure 13.1) from low to high latitude:

   a. along the west coast?

   b. in the midcontinent?

4. a. Plot T_B by latitude for the six west coast locations on the graph in Figure 13.5. Use a ruler to connect these points as short line segments. Label this the west coast transect. Then plot T_B by latitude for the midcontinent transect locations, and connect these points as a second set of short line segments. Label this the midcontinent transect.

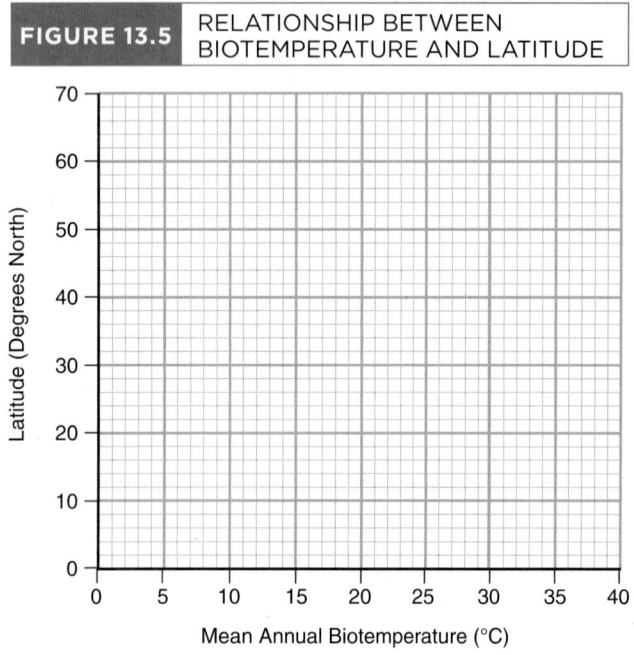

FIGURE 13.5 RELATIONSHIP BETWEEN BIOTEMPERATURE AND LATITUDE

   b. Plot Holdridge's ratio by latitude for the six west coast locations on the graph in Figure 13.6. Use a ruler to connect these points as short line segments. Label this as the west coast transect. Then plot Holdridge's ratio by latitude for the midcontinent transect locations and connect these points as a second set of short line segments. Label this the midcontinent transect.

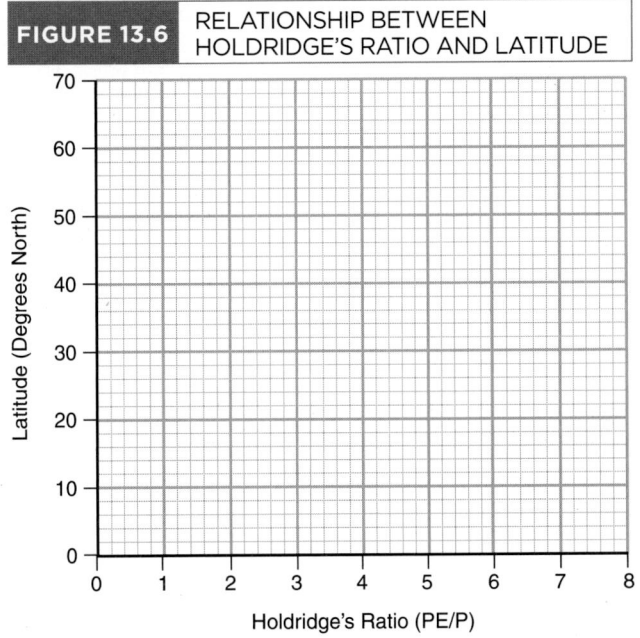

5. a. Calculate the mean annual biotemperature gradient across the 47° of latitude difference between Kugluktuk and Merida (the midcontinent transect) to the nearest 0.01. (Subscript 2 means the highest latitude site; 1 means the lowest.)

$$\frac{T_{B2} - T_{B1}}{\text{Latitude}_2 - \text{Latitude}_1} =$$

b. Calculate the mean annual biotemperature gradient across the 46° of latitude difference between Manzanillo and Fairbanks (the west coast transect) to the nearest 0.01.

$$\frac{T_{B2} - T_{B1}}{\text{Latitude}_2 - \text{Latitude}_1} =$$

c. Based on these two biotemperature gradients, what generalizations can you make about the change in biotemperature with latitude?

6. Two straight lines representing the biotemperature gradients for both latitudinal transects in Figure 13.5 would roughly parallel one another, but the west coast line would be slightly offset to the right from the midcontinent line.

a. What does the offset to the right signify as different about west coast biotemperatures compared to midcontinent biotemperatures?

b. Given that the west coast receives abundant precipitation due to ocean proximity and that biotemperature is calculated from months above water's freezing point, how might this difference in biotemperature affect vegetation communities on the west coast compared to the midcontinent at a given latitude?

c. The lines you drew should show that at about 40°N the biotemperature is higher in the midcontinent than along the west coast (unlike at other latitudes). What might account for the higher T_B (thus more favorable growing season) for the midcontinent transect site at this latitude?

7. a. Calculate the mean annual temperature gradient across the 47° of latitude difference between Kugluktuk and Merida (the midcontinent transect) to the nearest 0.01.

$$\frac{MAT_2 - MAT_1}{Latitude_2 - Latitude_1} =$$

b. Calculate the mean annual temperature gradient across the 46° of latitude difference between Manzanillo and Fairbanks (the west coast transect) to the nearest 0.01.

$$\frac{MAT_2 - MAT_1}{Latitude_2 - Latitude_1} =$$

c. What does the difference in the mean annual temperature gradient between the midcontinent and west coast latitudinal transects tell us?

d. How does the mean annual biotemperature gradient for the two latitudinal transects compare to the mean annual temperature gradient for the two latitudinal transects?

8. Based on the lines representing Holdridge's ratio on the graph in Figure 13.6, what generalization can you make about moisture conditions along the west coast compared to the midcontinent:

a. at latitudes greater than or equal to 40°N?

b. at latitudes less than 40°N?

c. What might account for the lower PE/P ratio (and thus moister growing conditions) for the west coast transect sites above 40°N, and what might account for the higher PE/P ratio (and thus drier conditions) at the west coast sites below 40°N?

d. How does moisture availability vary with latitude along the midcontinent transect?

9. Based on the information you've gathered on changes in biotemperature and Holdridge's ratio with latitude, and based on the sequences of expected changes in vegetation community with latitude (see Tables 13.2 and 13.5), which of these two variables (biotemperature or Holdridge's ratio) do you think plays a more important role in affecting changes in life zone with latitude:

   a. along the west coast?

   b. in the midcontinent?

10. Summarize the main similarities and differences between the west coast and the midcontinent transects:

    a. T_B

    b. PE/P

    c. life zones

11. a. How many of the site life zone community predictions were correct for the altitudinal transect (Table 13.3 and Figure 13.4)?

    b. List any erroneous sites.

    c. What might account for the mispredictions at any erroneous sites?

12. Figure 13.2 indicates that the altitudinal transect is located at about the same latitude as Concordia, which is part of the midcontinent latitudinal transect.

    a. Based on the life zone changes with latitude along the midcontinent transect, what changes in vegetation would you predict for the altitudinal transect, and at what altitudes would the changes occur (see Figure 13.4)?

    b. What is the *actual* sequence of vegetation change from low to high elevation along the altitudinal transect?

c. Comparing the vegetation changes along the midcontinent latitudinal transect and the altitudinal transect, does the change in vegetation in these two directions appear similar? Why would this be the case? Or, why would this not be the case?

13. a. Calculate to the nearest 0.0001 the mean annual temperature gradient across the 2243 meters of altitude difference between Niwot Ridge and Longmont.

$$\frac{MAT_2 - MAT_1}{Elevation_2 - Elevation_1} =$$

b. Calculate to the nearest 0.0001 the mean annual biotemperature gradient across the 2243 meters of altitude difference between Niwot Ridge and Longmont.

$$\frac{T_{B2} - T_{B1}}{Elevation_2 - Elevation_1} =$$

14. a. Divide the mean annual temperature gradient from the midcontinent latitudinal transect (question 7a) by that of the altitudinal transect. How many meters of altitude change have the same effect on MAT as increasing one degree of latitude in the midcontinent?

b. Divide the mean annual biotemperature gradient from the midcontinent latitudinal transect (question 5a) by that of the altitudinal transect. How many meters of altitude change have the same effect on T_B as increasing one degree of latitude in the midcontinent?

c. Figure 13.4 shows tundra beginning at 3400m. Use also, Figure 13.1, Table 13.5 data and your results from questions 14(a) and 14(b) to determine tundra elevation difference from Colorado to Williston. At what elevation on a hypothetical mountain in Williston would you reach tundra?

EXERCISE 14 † COINCIDENT CLIMATES, VEGETATION, AND SOILS

PURPOSE

The purpose of this exercise is to demonstrate the situational and site coincidences of various climates with particular types of vegetation and soil.

LEARNING OBJECTIVES

By the end of this exercise you should be able to

- explain the range of temperature and moisture conditions under which different biomes and soil orders exist;
- recognize the spatial correspondence of climatic conditions with particular vegetation and soils; and
- explain the respective importance of description, prediction, and explanation as objectives in science.

INTRODUCTION

Climatic conditions across the earth's surface vary in a somewhat predictable manner, and geographic distributions of both vegetation and soils conform approximately, but not perfectly, with climatic patterns. Among other things, organisms have adaptations to, and soils develop in response to, climate. As a result, particular organisms, and soils have typical associations with areas having suitable climates.

It is possible to describe geographical relationships within the natural environment by noting consistent coincidences of climates, organisms, and soils. **Description** (including classification), however, is only the first stage of scientific investigation.

Temperature and moisture are especially important climatic controls over the kinds of vegetation and soil that exist at a place, because these two climate elements strongly affect chemical reactivity. In general, warmer and/or wetter conditions tend to promote and accelerate the internal physiochemical processes of organisms and of biogeochemical reactions in soil. Cool and/or dry conditions usually inhibit such chemical activity. Ultimately, this basic relationship and the geographic coincidences of climate, vegetation, and soil provide some **explanation** for the natural landscapes that occur at various locations.

Prediction of the unknown spatial distributions of organisms, communities, and soils becomes possible by specifying known climatic conditions associated with similar organisms. Conversely, we can reasonably infer climates from the organisms and soils that exist at a place.

Representative Variables for Climate and Plant Growth

Since temperature and moisture are powerful influences over vegetation and soil development, you might expect that we would use these climatic factors as classification variables without any further modification. However, temperature is also the fundamental energy control over the processes that put the moisture into the air in the first place. Because of this linkage, it is more efficient to use variables that simultaneously represent both temperature and its consequent amount of atmospheric moisture at a place. In this exercise, those variables will be potential evapotranspiration (PE) and the index of moisture (Im).

The two main processes by which water vapor enters the atmosphere on earth are evaporation and transpiration. Other processes, such as sublimation, animal respiration, and volcanism are relatively minor contributors to atmospheric moisture. **Evaporation** occurs when the addition of heat energy to liquid water molecules causes the direct release of water vapor from a soil or water body. **Transpiration** is the release of water vapor to the atmosphere by green plants as a by-product of photosynthesis, the fixing of sunlight into carbohydrate chemical energy. **Evapotranspiration** is the combination of water going to the atmosphere from both evaporation and transpiration. Since the rates of both the vaporization of water and the growth of plants are partial functions of temperature, evapotranspiration is actually a composite variable that represents both temperature and moisture.

Potential evapotranspiration (PE) is the maximum amount of moisture that would evapotranspire if there were an unlimited supply of moisture. PE has a direct relationship to air temperature; as temperature increases, PE increases. In dry places, less water is transferred to the air than heat conditions might allow because very little water is available, and heat energy not used for evapotranspiration is used primarily to raise the temperature of the air. Thus, regardless of how much water is available, PE is an especially useful integrated climatic measure because it represents virtually all possible humidity at places where liquid water is abundant, yet also accounts for the excessive temperatures at dry places.

The Index of Moisture (Im)

The amount of moisture removed from the atmosphere as **precipitation (P)** varies from place to place. Depending on prevailing temperatures, the net moisture balance (total precipitation gains minus evapotranspiration losses) should be what is available to support plant growth and foster soil development. The size of this balance often is quite different from the absolute amount of precipitation received. A cold place with little precipitation may still have a positive moisture balance, while a rainy place that is very hot may run a perennial or seasonal moisture deficit. It is the net moisture *balance*—not the gross amount of moisture—at a place that helps control the form and quantity of vegetation and soil.

For comparative moisture classifications, an index value instead of an absolute amount standardizes the net moisture balance in a convenient ratio scale. This value is the **Index of moisture (Im)**, calculated as:

$$Im = 100[(P/PE) - 1] \quad (14.1)$$

where: P = mean annual precipitation (mm)
PE = potential evapotranspiration (mm).

Im values are dimensionless numbers, meaning they have no unit labels. The index magnitude is proportional to the moisture balance. Negative values indicate dry climates where P < PE. The more negative an Im value, the drier the climate. Positive values indicate wet climates where P > PE.

Vegetation and Soils

A **biome** is a characteristic assemblage of ecosystems, often associated with a particular climatic region. The five terrestrial ("land") biomes are forest, savanna, grassland, desert, and tundra. Within any biome there are more specific subsets of ecological communities called **formation groups**. The forest biome for example, includes tropical rainforest, various broadleaf temperate forests, and several needleleaf forest formation groups. The nonforest biomes, although they may have isolated trees, are open landscapes dominated by vegetation other than trees.

Soil orders are categories of the U.S. Department of Agriculture's organizational scheme for describing systematic characteristics and associations of soils. Twelve orders distinguish soils of various climatic, hydrologic, and geologic origins, and by their degree of development. As these same environmental conditions influence the vegetation communities that prevail at a place, it is quite common (although not absolute) that certain biomes and formation groups coincide with particular soil orders (Figure 14.1). For example, mollisols usually correspond with grassland or savanna, as these soils develop under the conditions of limited moisture that also preclude the existence of forest.

FIGURE 14.1 TEMPERATURE/MOISTURE GRAPHS

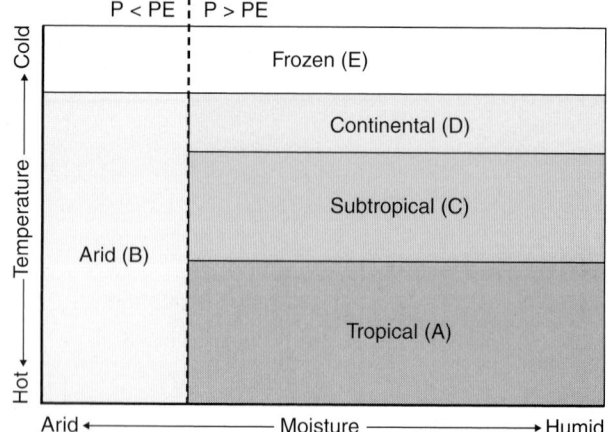

A. Climate

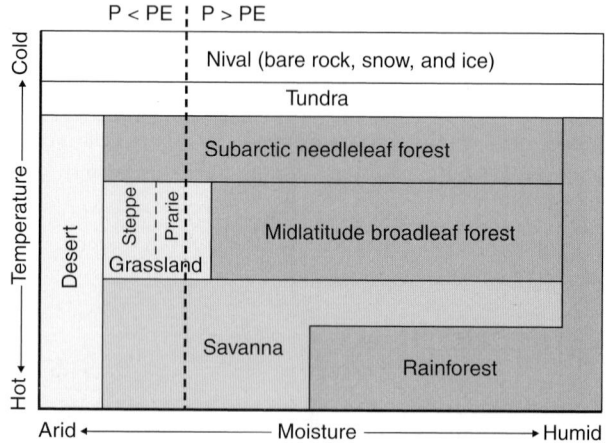

B. Vegetation

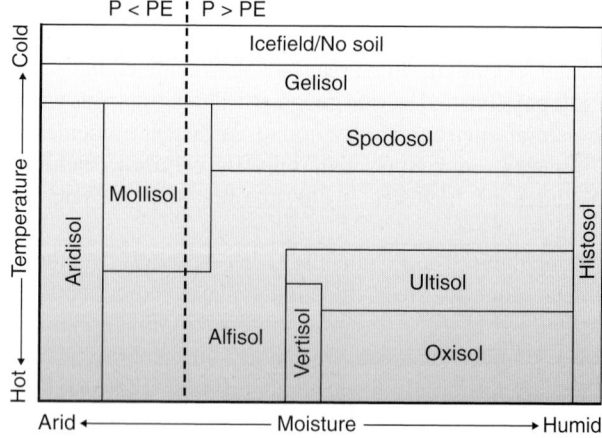

C. Soil

Source: After Gabler, et al., 1975.

Spatial Patterns

Systematic examination of climate, vegetation, and soils around the world reveals predictable coincidences between them, and regional associations of the resultant ecosystems. The consistency of these patterns suggests that climate is a strong underlying control that influences jointly the spatial placement of vegetation and soil (Appendix F, Figures F.3 and F.4). The exercise that follows is a small-scale reenactment of an original study by Mather and Yashioka (1968), which used data from thousands of sites worldwide. Much of the discrepancy in detail between the exercise graph patterns and those of Figure 14.1 results from the use of only 50 data sites here, but certain regular coincidences between climate, vegetation, and soil types will begin to emerge even from just these few exercise locations.

IMPORTANT TERMS, PHRASES, AND CONCEPTS

description
explanation
prediction
evaporation
transpiration
evapotranspiration
potential evapotranspiration
precipitation
index of moisture
biome
formation group
soil order

REFERENCE

Mather, J.R. and G.A. Yoshioka. 1968. The role of climate in the distribution of vegetation. *Annals of the Association of American Geographers* 58: 29–41.

Name: _____   Section: _____

PART 1 † DATA COMPILATION AND CLASSIFICATION

1. Calculate Im to the nearest whole number for the last ten stations in Table 14.2 using equation 14.1.
2. Use latitude and longitude to locate these same ten stations on world maps of vegetation (Appendix E, Figures E.1 and E.3 and Appendix F, Figure F.3) and soils (Appendix F, Figure F.4) to determine their vegetation and soil. To complete Table 14.2, record a letter code representing the vegetation and a letter code for soil at each station using the letters in Table 14.1. Tropical rainforest will be forest on Figure F.3, broadleaf on Figure E.1, and evergreen on Figure E.3. Broadleaf temperate forest will be forest on Figure F.3, broadleaf or mixed on Figure E.1, and deciduous on Figure E.3. Needleleaf forest will be forest on Figure F.3, and needleleaf on Figure E.1.

TABLE 14.1 — VEGETATION AND SOIL LETTER CODES

Use:	For:	Use:	For:
R	Tropical rainforest	s	spodosols ("ash-colored")
B	Broadleaf temperate forest	a	alfisols ("aluminum and iron stains")
N	Needleleaf forest	m	mollisols ("soft and friable")
S	Savanna	u	ultisols ("old")
G	Grassland	o	oxisols ("oxidized")
D	Desert	v	vertisols ("turning over")
T	Tundra	h	histosols ("tissue-like")
		d	aridisols ("dry")
		e	entisols ("recent")
		i	inceptisols ("beginning")
		n	andisols ("volcanic andesite")
		g	gelisols ("gelifluction and frost heaving")
		x	highland soils (use first listed, if available)

TABLE 14.2 — STATION DATA

#	Station	Location	P (mm)	PE (mm)	Im	Vegetation	Soil
1	Buenos Aires, Argentina	34°S, 59°W	1005	759	+32	G	m
2	Belem, Brazil	1°N, 48°W	2833	1537	+84	R	i

(continued)

167

TABLE 14.2 STATION DATA (CONTINUED)

#	Station	Location	P (mm)	PE (mm)	Im	Vegetation	Soil
3	Monterrey, Nuevo Leon	26°N,100°W	606	1100	−45	D	d
4	Oklahoma City, Oklahoma	35°N,98°W	818	776	+5	G	m
5	Tucson, Arizona	32°N,111°W	295	1024	−71	D	d
6	El Oregano, Sonora	29°N,111°W	363	1731	−79	D	d
7	Tonopah, Nevada	38°N,117°W	134	576	−77	D	d
8	Berthoud Pass, Colorado	40°N,106°W	958	285	+236	T	x
9	Williston, North Dakota	48°N,104°W	352	487	−28	G	m
10	Minneapolis, Minnesota	45°N,93°W	701	532	+32	B	a
11	Fairbanks, Alaska	65°N,148°W	268	357	−25	N	g
12	Barrow, Alaska	71°N,156°W	112	222	−49	T	g
13	Hay River, Northwest Territories	60°N,117°W	326	336	−3	N	s
14	Cartwright, Newfoundland	54°N,57°W	966	336	+187	T	e
15	Dryden, Ontario	50°N,93°W	671	424	+58	N	s
16	Moose Jaw, Saskatchewan	50°N,106°W	368	453	−19	G	m
17	Edmonton, Alberta	54°N,113°W	454	426	+7	N	a
18	Windsor, Ontario	42°N,83°W	831	549	+51	B	a
19	Churchill, Manitoba	59°N,94°W	402	268	+50	T	h
20	Reykjavik, Iceland	64°N,22°W	818	410	+100	T	g
21	Cork, Ireland	52°N,8°W	1081	563	+92	B	a
22	Turku, Finland	60°N,22°E	632	419	+51	B	s
23	Murmansk, Russia	69°N,33°E	412	321	+28	N	s
24	Irkutsk, Russia	52°N,104°E	418	378	+11	N	i
25	Urumqi, China	44°N,88°E	254	558	−54	D	d
26	Ho Chi Minh City, Vietnam	11°N,107°E	1934	1712	+13	R	u
27	Kolkata, India	23°N,88°E	1633	1781	−8	S	u

(continued)

TABLE 14.2	STATION DATA *(CONTINUED)*						
#	Station	Location	P (mm)	PE (mm)	Im	Vegetation	Soil
28	Tehran, Iran	36°N,52°E	239	848	−72	D	d
29	Singapore, Singapore	1°N,104°E	2272	1687	+35	R	o
30	Cairo, Egypt	30°N,31°E	28	1035	−97	D	d
31	Douala, Camaroon	4°N,10°E	4000	1380	+165	R	o
32	Windhoek, Namibia	22°S,17°E	367	865	−58	D	d
33	Cape Town, South Africa	34°S,18°E	611	753	−19	S	a
34	Mombasa, Kenya	4°S,40°E	1144	1587	−28	S	a
35	Entebbe, Uganda	0°,32°E	1535	971	+58	S	u
36	Banjarmasin, Indonesia	3°S,115°E	2553	1610	+59	R	o
37	Maputo, Mozambique	27°S,33°E	767	1078	−29	S	d
38	Kisangani, Democratic Republic of the Congo	1°N,25°E	1858	1273	+46	R	o
39	Harare, Zimbabwe	18°S,31°E	827	813	+2	S	a
40	Luanda, Angola	9°S,13°E	369	1207	−69	S	d
41	Green Bay, Wisconsin	45°N,88°W	725	504			
42	Longmont, Colorado	40°N,105°W	327	544			
43	St. Louis, Missouri	39°N,90°W	939	698			
44	Washington, D. C.	39°N,77°W	990	725			
45	Atlanta, Georgia	34°N,84°W	1237	783			
46	Manaus, Brazil	3°S,60°W	2081	1611			
47	Iquitos, Peru	4°S,73°W	2878	1519			
48	Leticia, Colombia	4°S,70°W	2811	884			
49	Iquique, Chile	20°S,70°W	2	793			
50	Brasilia, Brazil	16°S,48°W	1555	920			

3. Figures 14.2 (vegetation graph) and 14.3 (soils graph) were created using the data in Table 14.2. Plot the vegetation and soil letter for the last ten stations on each graph.

FIGURE 14.2 VEGETATION GRAPH

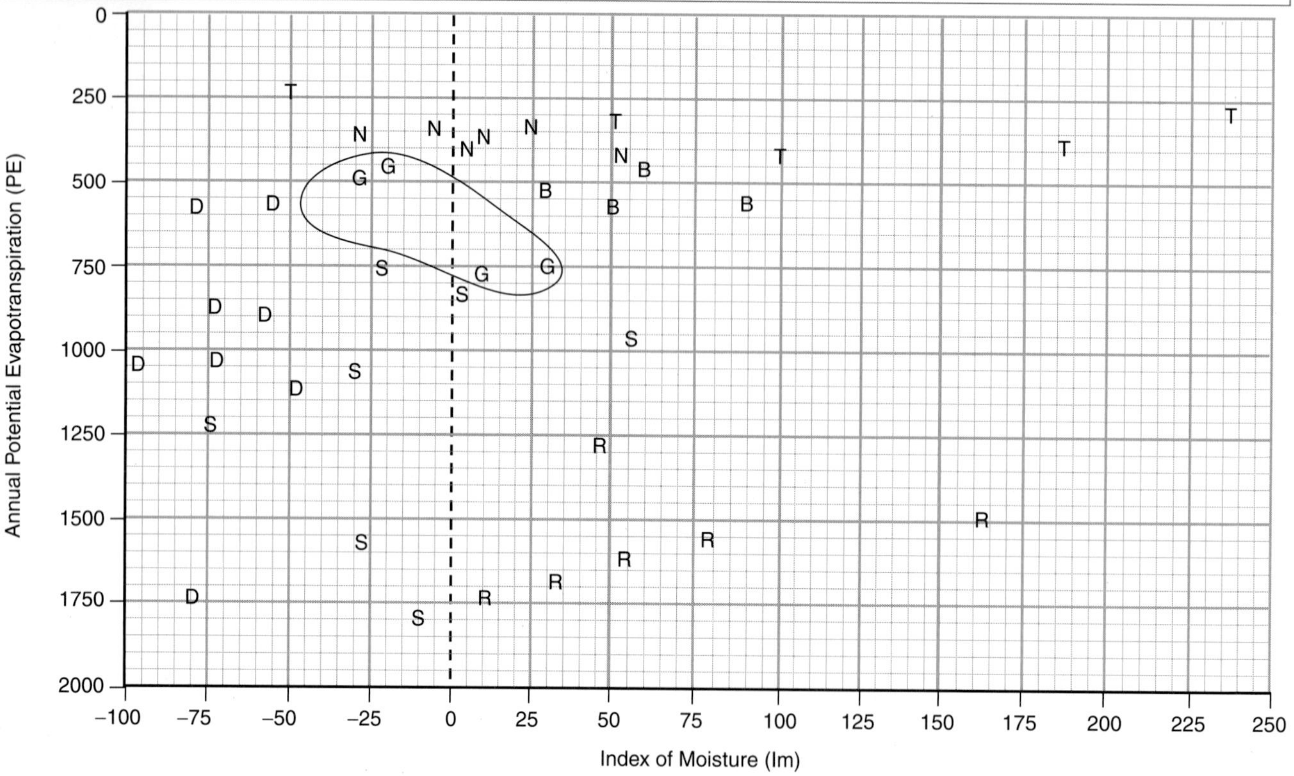

FIGURE 14.3 SOIL ORDERS GRAPH

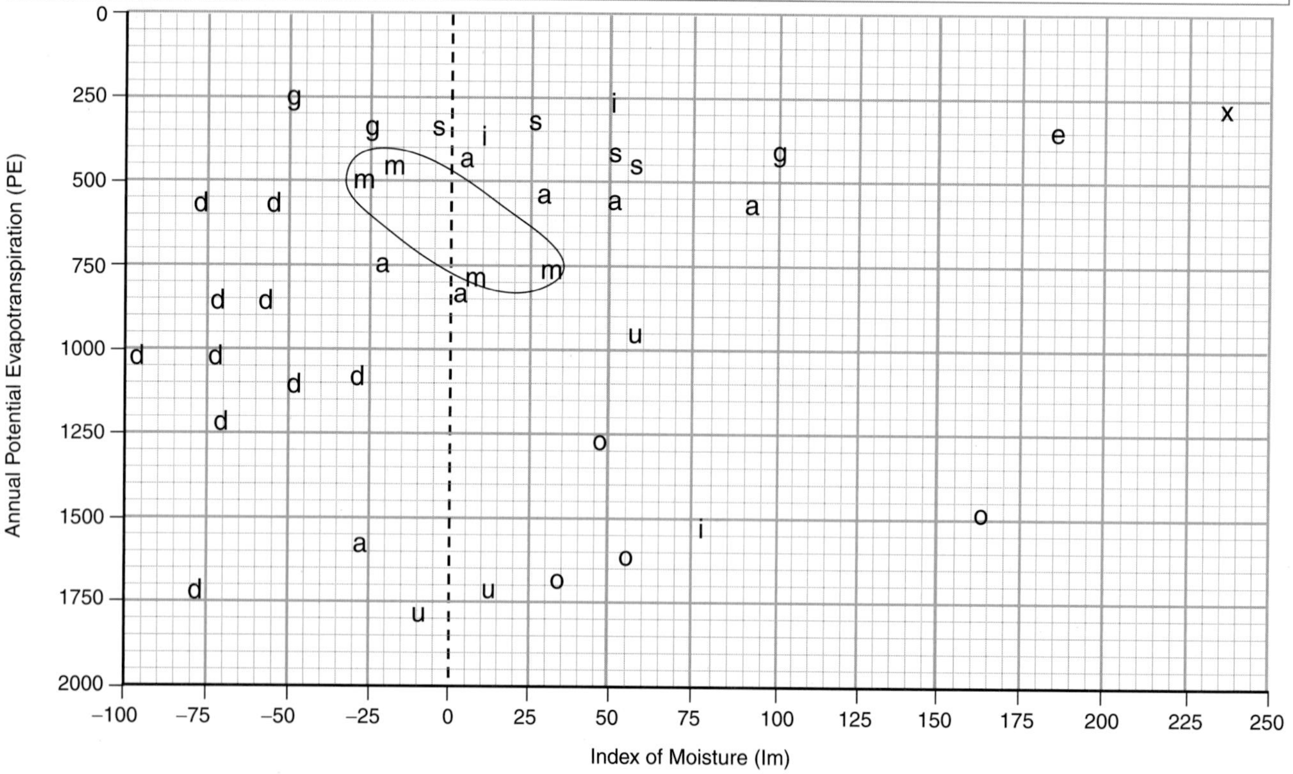

170

4. Using grassland (G) and mollisol (m) as examples, encircle clusters of the same letters on the vegetation graph (Figure 14.2) and the soils graph (Figure 14.3) except for entisols (e) and inceptisols (i). These enclosures indicate the variety of temperature and moisture conditions with which a particular vegetation or soil typically coincides. The enclosure lines may overlap, indicating multiple coincidences. Shade the enclosures in different colors to make the patterns clear. It would be a good idea to do this in pencil!

5. Compare your vegetation graph and your soils graph with Figure 14.1. While you will have some difference in appearance, there should also be some degree of correspondence between your patterns and those of Figure 14.1.

   a. Which of your vegetation and soil zones seem to match Figure 14.1 the best?

   b. Which of your vegetation and soil zones seem to match Figure 14.1 the least?

   c. What might explain the imperfect correspondence between your graphs (Figures 14.2 and 14.3) and Figure 14.1?

6. List all of the vegetation types and soil orders from your vegetation graph (Figure 14.2) and your soils graph (Figure 14.3) that correspond with the following climates:

	Vegetation Letter Code(s)	Soil Letter Code(s)
Tropical (Köppen A)	_____	_____
Subtropical (Köppen C)	_____	_____
Continental (Köppen D)	_____	_____
Frozen (Köppen E)	_____	_____
Arid (Köppen B)	_____	_____

Name: _____   Section: _____

PART 2 † INTERPRETIVE APPLICATIONS

DESCRIPTION

1. What soil(s) and climate(s) best correspond to each vegetation? Use Table 14.2, Figure 14.2 (vegetation graph), and Figure 14.3 (soils graph).

	Soil Code	Climate (e.g., hot/cold; wet/dry)
Tropical rainforest (R)	_____	_____
Broadleaf temperate forest (B)	_____	_____
Needleleaf forest (N)	_____	_____
Savanna (S)	_____	_____
Grassland (G)	_____	_____
Desert (D)	_____	_____
Tundra (T)	_____	_____

2. Draw a line that separates forest from nonforest vegetation on the vegetation graph (Figure 14.2). This line shows the environmental limits to forest vegetation, something we can actually see on the landscape—the treeline: a boundary beyond which trees do not grow.

3. a. What would be a good approximate cutoff Im value separating savanna from forest?

    b. What would be a good approximate cutoff PE value separating tundra from forest?

    c. By your cutoff values, describe the temperature and moisture conditions necessary for forest to exist.

EXPLANATION

Some soil orders (e.g., aridisols) cluster pretty much into their own exclusive area on Figure 14.3. Other orders (e.g., alfisols) seem to share part of their graph zones with other soils. And some orders, such as inceptisols and entisols, do not seem to cluster very much at all, but appear almost randomly on the graph. Consider the meaning of the axis variables in Figures 14.2 and 14.3 as you try to explain soil occurrence tendencies.

4. Name a soil order other than aridisol that clusters into its own graph zone. What vegetation (on Figure 14.2) does this soil consistently coincide with, and what climatic conditions evidently exist where this soil and vegetation develop?

Soil	Vegetation	Climate
_____	_____	_____

5. a. Name two soil orders that overlap and thus share a graph zone.

    b. What climatic conditions (e.g., hot/cold, wet/dry) does their overlap represent?

173

c. What nonclimatic factors might cause these two different soils to form under similar climatic conditions?

6. What nonclimatic situations could produce entisols and inceptisols across a variety of different climates, as appears evident within Figure 14.3?

7. Some of the vegetation (Figure 14.2) may overlap or contain outliers, implying that several types of vegetation coexist together. Give one possible explanation for joint (instead of exclusive) occupation of the same graph zone by unlike vegetation.

PREDICTION

8. Using your graphs and comparing the PE and Im values in Table 14.1, what vegetation and soil (use letter codes from Table 14.1) do you predict for the following hypothetical places (Table 14.3)? Then indicate locations where each of these ecosystems might actually occur.

TABLE 14.3			CONDITIONS IN NORTH AMERICA		
PE	Im	Vegetation	Soil	Likely North America Location	
375	−5				
610	−31				
733	+27				
1225	−95				
1475	+105				

EXERCISE 15 ✝ HAWAI'I RAINFOREST REGENERATION

PURPOSE

The purpose of this exercise is to assess change in a Hawaiian rainforest, as evident in sequential aerial imagery, following catastrophic volcanic disturbance to its plant cover.

LEARNING OBJECTIVES

By the end of this exercise you should be able to

- use aerial imagery to analyze rainforest cover within a small area;
- describe and compare current habitat conditions at different places;
- interpret the qualities, quantities, and changes in vegetation evident in sequential images; and
- assess habitat adjustments that occur through time at disturbed locations.

INTRODUCTION

Geographic study entails not simply the description or comparison of current conditions at different places but also the assessment of changes that occur through time at single locations. This lab introduces some characteristics and applications of aerial imagery for analyzing rainforest cover through interpretation of changes evident in sequential images within a small area.

A **forest** by definition is an area under a continuous canopy (meaning that the foliage of one individual physically contacts the foliage of several neighbors) of tree-form plants (over 1.5 meters tall and having a woody stem). Many kinds of forests exist, but even within a rainforest continuous tree cover usually does not occupy all ground. Rainforest normally has dense vegetation, growing year-round in hot, humid climates, with quite a diverse array of co-existing plant and animal species.

To be ecologically healthy and maintain high diversity, however, rainforests also need local patches of nonforest community as special habitats (e.g., wetlands, disturbance gaps) within them. **Ecotones** are habitat transition zones and are rather sensitive to and thus good indicators of environmental changes and disturbance. Forests exhibit short-term ecotone changes, during which the ecotones undergo a predictable sequence of changes in composition and form. This sequence of changes is called **succession.** Lasting redistributions of vegetation also may occur and often reflect a long-term shift in some influential environmental quality, such as climate.

Hawaii Volcanoes National Park: the Devastation Trail Study Area

Kilauea Iki on Hawai'i (Figure 15.1) vigorously erupted between 14 November and 20 December 1959 in Hawaii Volcanoes National Park, creating a natural ecotone between intact Hawaiian rainforest toward the east and damaged rainforest toward the west. Fumes and ashfall from the eruption caused an instantaneous—but also quite localized—**disturbance.** Additional ecotones reflect further wildfire, insect infestation, land clearing, and selective planting disturbances.

Some fifty years following the 1959 eruption disturbance, the recovery process at Kilauea Iki remains incomplete; full regeneration of Hawaiian rainforest can take centuries or millennia. *Metrosideros polymorpha* ('ohi'a lehua tree), a common tree species in Hawaiian rainforest, is dispersing now into post-eruption ecotones but still remains smaller in the damaged area than in intact rainforest. 'Ohi'a lehua can be a light-loving, fast-growing pioneer species of Hawai'i lava flows and ash fields, but the typical growth rate for mature individuals in rainforest is 0.15 m/yr. Unlike the 'ohi'a lehua tree, another common Hawaiian rainforest species, *Cibotium glaucum* (hapu'u pulu tree fern), is still relatively uncommon in the damaged area, as revealed in a 2006 site visit. Those few hapu'u pulu tree ferns present within the **damaged rainforest** habitat were mere sprouts; mature individuals were tree-size only within intact rainforest. Other kinds and numbers of plants and animals also have occupied, or will occupy, the damaged rainforest during succession phases as the area gradually returns to its original condition.

Disturbance and succession alter an ecosystem's **carrying capacity,** the maximum number of individuals that a habitat can support. Some habitat conditions may initially favor one species with high carrying capacity but subsequently yield to phases that benefit some other species more. Population changes of two native Hawaiian birds illustrate this. The endangered 'io (Hawaiian hawk, *Buteo solitarius*) fares better in the open habitat of damaged rainforest where its acute eyesight enables greater hunting success. Initially, its

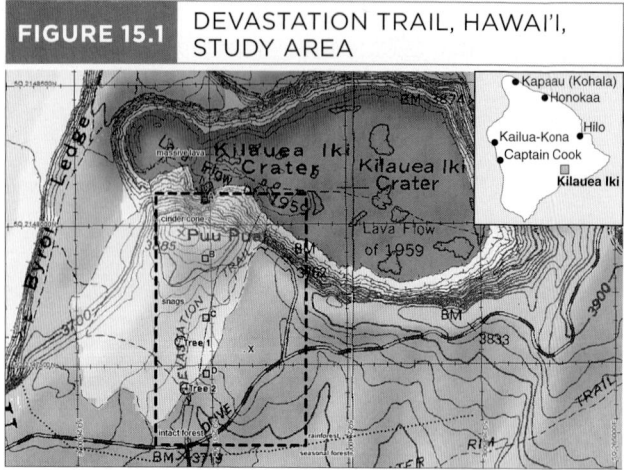

FIGURE 15.1 DEVASTATION TRAIL, HAWAI'I, STUDY AREA

Source: Map Courtesy of the USGS.

population increased after the eruption but then declined with forest regrowth. The Hawai'i o'o (a honeycreeper, *Moho nobilis*) suffered severe population decline when its habitat and 'ohi'a lehua blossom nectar supplies diminished but later might have experienced greater carrying capacity with regeneration of additional ohi'a lehua forest habitat. Sadly, we will never know if it could have recovered; the Hawai'i o'o became extinct due to habitat loss.

In ecosystem studies, **quadrats,** small square areas of a consistent size, allow scientists to select and study representative tracts of land. This enables comparison of different ecosystems, or parts of a single ecosystem, without having to examine the entire study area. Four quadrats are marked on Figure E.5 (Appendix E), and each represents a distinct habitat (Smathers and Mueller-Dombois, 1974) that appears in the aerial images for this exercise.

The first vegetation habitat, a massive **lava flow** (quadrat A), was initially barren, but even by 1960 a few pioneer plants had colonized from the upwind (northeasterly) rainforest. In 1987 (the date of Figure E.5) and in 2006, however, the lava flows at quadrat A still contained very few plants.

Some pioneer rainforest species re-seeded the second vegetation habitat, the Pu'u Puai ("fountain hill") **cinder cone** habitat illustrated within quadrat B, but the aridity of the loose ash soils here precluded survival of most species except very drought-tolerant, exotic (nonnative) ones. By the time of a 2006 site visit, this quadrat was again barren.

The third vegetation habitat, exemplified by quadrat C, contained snags, which are standing, dead tree trunks. This **snag** habitat included an ecotone upwind from the cinder cone, where a few volcano-adapted rainforest plants survived tand resprouted, but most plants farther downwind succumbed to the deep (up to 12 meters) burial by hot ash. By 1987 many of the snags had fallen, and in 2006 relatively few remained standing.

Finally, the fourth vegetation habitat was **intact rainforest** (quadrat D). This was more distant from Pu'u Puai and so less deeply buried, thus enabling survival of most vegetation. Its condition in 2006 was much the same as before the eruption, except for the nearby addition of Devastation Trail for park visitors.

Habitat and Plant Identification

Scientists, such as biogeographers and foresters, routinely investigate forest ecosystems through analyses of aerial imagery, ranging from inventories of species to assessment of ecosystem change processes. Such analyses can hasten investigations and hold down costs by reducing the need for laborious ground surveys. However, analysts should ground-truth (spot check during a site visit) their interpretations from image analyses whenever possible.

Trained professionals can identify and map vegetation, even to the species level of taxonomy, from qualities evident on aerial images. The presence—or absence—of these various image qualities helps enable the differentiation of vegetation forms, and these qualities should always be considered in conjunction with one another, not singly. Vegetation interpretation can become ambiguous due to the normal variability of form between individuals of the same kind. To identify plants within a stand, examine multiple individuals and seek a dominant tendency among them for each interpretive quality. The image qualities that we will use in this exercise include:

- **Texture:** the roughness of an object's image tones (e.g., coarse, dappled; smooth, uniform). In Figure E.5, the surface around quadrat B is noticeably smoother than around the other quadrats; quadrat B is a surface of primarily fine ash deposits, and virtually no vegetation. At the grass area toward the west (left) of the north parking lot, the image texture is clearly much smoother than in adjacent forest.

- **Shadow:** the presence, shape, relative size, and/or pattern of shadows cast by vertical object(s) in oblique illumination, but not the characteristics of an object itself. In Figure E.5, there is no shadow for the low shrub at the north parking lot or at the grass area, but close examination reveals a full tri-lobate black shadow for trees at the south parking lot. Likewise, tree 1 (Appendix E, Figure E.5 and Figure 15.2) and tree 2 (Appendix E, Figure E.5 and Figure 15.3) have quite different shadow lengths, shapes, and fullness.

- **Tone:** the color (green, red, etc.) and/or the grey-scale intensity (bright, light; dull, dark) of an object. The imagery for this exercise is "false-color infrared," where what prints as red was most reflective in invisible near infrared (NIR) light. Green in the images actually was visual red, and image blue was green to the human eye. This is a useful color scheme, because healthy, thriving vegetation is more reflective ("peaks") in NIR rather than in colors visible to the human eye, so it appears red in false-color. Other examples of surface material appearances in false-color infrared imagery are listed in Table 15.1.

| FIGURE 15.2 | TREE 1 SHADOW ENLARGEMENT |

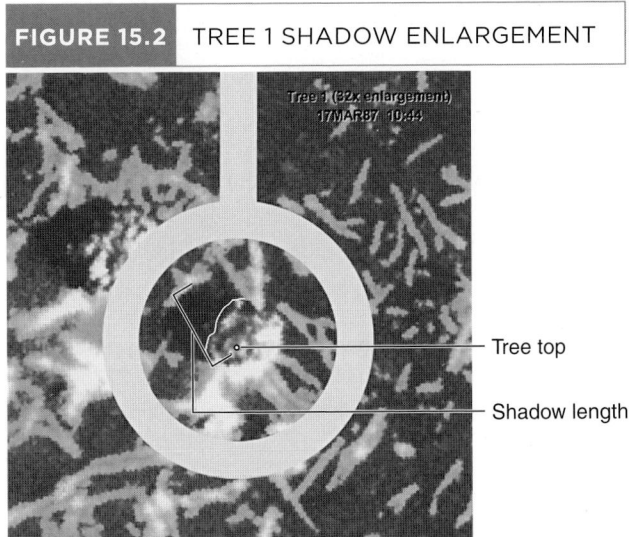

Source: Courtesy of NASA Ames Research Center.

| FIGURE 15.3 | TREE 2 SHADOW ENLARGEMENT |

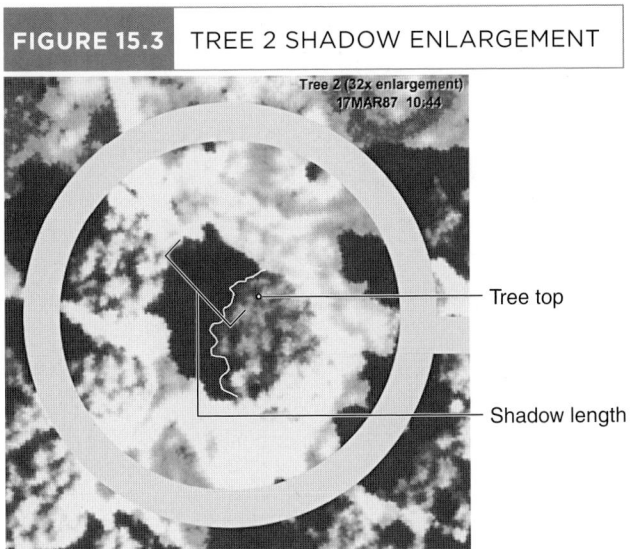

Source: Courtesy of NASA Ames Research Center.

Intact rainforest is apparent as image red throughout the east side of Figure E.5. Toward the west from quadrat A appears a greenish area; this is iron-rich red rock, barren of vegetation, at the Pu'u Puai vent. Stressed or dormant vegetation appears greenish blue around tree 1, while the barren ash downwind from Pu'u Puai and some roads are bluish grey. At quadrat A lava flows appear rather darker than ash and roads.

Aerial images are preservable records of conditions at different times and allow us to reconstruct past conditions and document changes of habitat over time through analysis of archival imagery. By measuring the same qualities at a place for successive dates, it becomes possible to determine adjustments in ecosystem composition, carrying capacity shifts, and rates of landscape change.

| TABLE 15.1 | TONE INTERPRETATION ON FALSE-COLOR INFRARED IMAGERY |

Actual Peak Color	Image Color	Example Materials
NIR	red	growing vegetation
red	green	red autumn foliage, iron-rich rock or soil
brown	greenish blue	stressed vegetation; dark, moist soil or rock
green	blue	dormant evergreen vegetation, sea surf
blue, grey, or white	bluish grey	dry soil, pumice ash, concrete, ice
black	black	barren basalt flows, asphalt, open water

Scale Conversion

Like maps, vertical-perspective aerial images have a **scale**, which is the proportional reduction between the distance shown on an image and the actual distance on the ground; think of it as a "shrinkage" factor from reality to paper. A useful way to express image scale is as a **representative fraction** (a ratio), such as 1:1000. Such a ratio means that one unit of linear distance on the image (the numerator, which is always "1") represents 1000 of the same units on the ground. Any unit of measurement may be used with a representative fraction, just so long as the units are the same during initial interpretation.

It is quite rare, however, that image objects are conveniently at an exact one distance unit apart, so it becomes necessary to convert actual distance measurements from the image into ground distances. As an example, the median barrier in the south parking lot of Figure E.5 (west of the tree with the shadow) is 75 meters long in reality, but appears only 32.258 mm long on these 1:2325 images. The representative fraction applies directly to linear measurements only; to convert areas, you must square the representative fraction denominator.

The linear ground distance of the parking barrier in kilometers therefore is:

$$X \text{ km} = \frac{32.258 \text{ mm}_{image}}{1} \times \frac{2325 \text{ mm}}{1 \text{ mm}_{image}} \times \frac{1 \text{ m}}{1000 \text{ mm}} \times \frac{1 \text{ km}}{1000 \text{ m}} = 0.075 \text{ km} \quad (15.1)$$

If the area of the south parking lot on this image were 1956 mm², the ground area would be:

$$X \text{ km}^2 = \frac{1956 \text{ mm}^2_{image}}{1} \times \frac{5,405,625 \text{ mm}^2}{1 \text{ mm}^2_{image}} \times \frac{1 \text{ m}^2}{1,000,000 \text{ mm}^2} \times \frac{1 \text{ km}^2}{1,000,000 \text{ m}^2}$$

$$= 0.0106 \text{ km}^2 \quad (15.2)$$

In addition to measuring lengths or areas on an image, we may also want to measure the height of various plants. Height comparisons between individuals of the same species often are indicative of relative age and growth. The purpose of making height measurements might be either to compare the relative ages of different stands on the same image or to assess the amount of growth that has occurred within the same stand between successive image dates.

Under proper conditions, the height of a single object may be calculated using its shadow length on an aerial image (Figure 15.4), then converted to ground length. The image height (H) of a tree is the image length of its shadow (L) multiplied times the tangent of the sun angle (s) at the exact moment of the image, as calculated by the equation:

$$H = L(\tan s) \qquad (15.3)$$

where: H = tree height
L = shadow length
s = sun angle

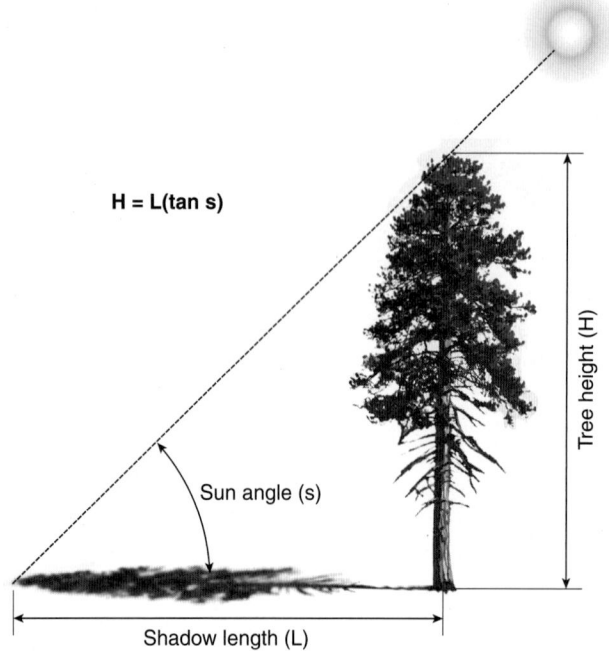

FIGURE 15.4 HEIGHT DETERMINATION FROM SHADOW LENGTH

This technique, however, provides accurate heights only when:

1. A vertical tree casts its farthest shadow from its true top.
2. The shadow is discernable from other shadows, and falls on level ground.
3. The tree's latitude, longitude, date, and time are available (for calculating sun angle).

Suppose that the shadow of an 'ohi'a lehua tree fell onto fresh pumice ash and measured 2 mm on a 1:2325 image where the sun angle was 35°. The image height of this tree would be:

$$H = 2 \text{ mm} \times (\tan 35°) = 1.4 \text{ mm}_{image}$$

This image height would then convert into ground height just like any other linear distance:

$$X \text{ m} = \frac{1.4 \text{ mm}_{image}}{1} \times \frac{2325 \text{ mm}}{1 \text{ mm}_{image}} \times \frac{1 \text{ m}}{1000 \text{ mm}} = 3.255 \text{ m}$$

IMPORTANT TERMS, PHRASES, AND CONCEPTS

forest	texture
ecotone	shadow
tone	quadrats
damaged rainforest	intact rainforest
representative fraction	snag
carrying capacity	cinder cone
succession	lava flow
disturbance	scale

REFERENCE

Smathers, Garret A. and Dieter Mueller-Dombois. 1974. *Invasion and Recovery of Vegetation after a Volcanic Eruption in Hawaii*. National Park Service. Scientific Monograph Series, no. 5. Publication number NPS 118. Supt. of Docs. no. I29.80:5.

Name: _____   Section: _____

PART 1 † FEATURE AND HABITAT IDENTIFICATION

Appendix E, Figure E.5 shows Devastation Trail rainforest on 17 March 1987, some 27 years after the eruption.

1. Describe the image qualities for each of the quadrats.

	habitat	tone (both color and intensity)	texture	shadow presence or absence
Quadrat A	lava flow	_____	_____	_____
Quadrat B	cinder cone	_____	_____	_____
Quadrat C	snag	_____	_____	_____
Quadrat D	intact rainforest	_____	_____	_____

2. Based on shadow presence and shadow length, what plant forms (tree, shrub, grass, or none) appear in each quadrat?

Quadrat A _____   Quadrat B _____

Quadrat C _____   Quadrat D _____

3. Suppose that a new eruption in March 1987 had created a cinder cone at the road junction on the image's east side (none actually did). Given the prevailing northeasterly winds on Hawai'i, what would you expect to see within each quadrat on a hypothetical June 1987 image?

	June 1987 habitat	June 1987 tone (both color and intensity)	June 1987 texture	June 1987 shadows
Quadrat A	_____	_____	_____	_____
Quadrat B	_____	_____	_____	_____
Quadrat C	_____	_____	_____	_____
Quadrat D	_____	_____	_____	_____

4. a. Describe the interpretation qualities for each quadrat on the 2001 image (Appendix E, Figure E.6).

	tone (both color and intensity)	texture	shadow presence or absence
Quadrat A	_____	_____	_____
Quadrat B	_____	_____	_____
Quadrat C	_____	_____	_____
Quadrat D	_____	_____	_____

b. Based on the image qualities you listed above, describe how each of these habitats has changed since 1987 and justify your description.

Quadrat A:

Quadrat B:

179

Quadrat C:

Quadrat D:

5. a. Compare the appearance of the four quadrats in the 1987 image (Figure E.5) with the same quadrats marked on the 2001 image (Figure E.6). One of them changed texture and tone most dramatically; which quadrat is this?

   b. The area within this quadrat apparently changed from what habitat in 1987 to what habitat in 2001?

Name: _____   Section: _____

PART 2 † ASSESSING HABITAT CHANGES

The large rectangular study area perimeter on the 1987 (Figure E.5) and 2001 (Figure E.6) images encloses the exact same tract. Close examination of intact and damaged rainforests in these images reveals quite a bit of change that occurred over this 26-year period. So, "How much change, of what, and with what implications?"

The following data derive from 1974 (image not shown), 1987, and 2001 aerial imagery.

Local Date	Local Time	Sun Angle	Tangent	Scale
19 October 1974	10:35 AM	53.0°	1.32704	1:8654
17 March 1987	10:44 AM	56.9°	1.53400	1:2325
21 November 2001	10:55 AM	49.2°	1.15851	1:2325

Because small-distance measuring instrumentation is not available to you, we provide the image measurements of tree shadow lengths (in image millimeters) for making height calculations of Trees 1 and 2. Both trees are 'ohi'a lehua.

As an example, the shadow lengths for Trees 1 and 2 in 1974 (this image is not shown) were 0.057 mm and 0.138 mm, respectively. Using equation 15.2 for Tree 1 in 1974, we get:

$$H_{image} = 0.057 \text{ mm} \times (\tan 53°) = 0.076 \text{ mm}_{image}$$

as the image height of the tree. Using equation 15.1 to convert this image height to a ground height, we get:

$$X \text{ m} = \frac{0.076 \text{ mm}_{image}}{1} \times \frac{8654 \text{ mm}}{1 \text{ mm}_{image}} \times \frac{1 \text{ m}}{1000 \text{ mm}} = 0.66 \text{ m}$$

Thus the actual height of Tree 1 was 0.66 meters. Then, using equation 15.2 for Tree 2 in 1974, we get:

$$H_{image} = 0.138 \text{ mm} \times (\tan 53°) = 0.18 \text{ mm}_{image}$$

as the image height of the tree. Again using equation 15.1 to convert this image height to a ground height, we get:

$$X \text{ m} = \frac{0.18 \text{ mm}_{image}}{1} \times \frac{8654 \text{ mm}}{1 \text{ mm}_{image}} \times \frac{1 \text{ m}}{1000 \text{ mm}} = 1.56 \text{ m}$$

Thus the actual height of Tree 2 is 1.56 meters.

1. Look at the shadows (dark tone) in the 1987 aerial image enlargements (Figures 15.2 and 15.3). Does Tree 2 appear to have a longer shadow than Tree 1?

2. How tall were these two trees, in meters, by 1987 (scale = 1:2325)?

   Tree 1: shadow length $_{image}$ = 0.676 mm

   Tree 2: shadow length $_{image}$ = 1.372 mm

3. How tall were these two trees, in meters, by 2001?

   Tree 1: shadow length $_{image}$ = 2.357 mm

Tree 2: shadow length $_{image}$ = 3.152 mm

4. a. How much vertical growth occurred for each circled tree over the 27 years?

   Tree 1: _____   Tree 2: _____

   b. How did growth for each tree deviate from the normal 'ohi'a lehua growth rate of 0.15 meters per year?

   c. What might account for the departure from normal growth rate?

The total ground area within the study quadrat is 0.5 km². Eruption-damaged forest includes the lava flow (quadrat A), cinder cone (quadrat B), and snag (quadrat C) habitats. Determine the damaged and undamaged forest area within the study area. Remember to square the representative fraction denominator to convert areas. As an example, the 1974 calculations are:

The total study area on the 1974 image (not shown) was 6676 mm², and the image scale was 1:8654.

$$X \text{ km}^2 = \frac{6676 \text{ mm}^2_{image}}{1} \times \frac{74{,}891{,}716 \text{ mm}^2}{1 \text{ mm}^2_{image}} \times \frac{1 \text{ m}^2}{1{,}000{,}000 \text{ mm}^2} \times \frac{1 \text{ km}^2}{1{,}000{,}000 \text{ m}^2} = 0.500 \text{ km}^2$$

The 1974 image area of the eruption-damaged rainforest was 3365 mm²:

$$X \text{ km}^2 = \frac{3365 \text{ mm}^2_{image}}{1} \times \frac{74{,}891{,}716 \text{ mm}^2}{1 \text{ mm}^2_{image}} \times \frac{1 \text{ m}^2}{1{,}000{,}000 \text{ mm}^2} \times \frac{1 \text{ km}^2}{1{,}000{,}000 \text{ m}^2} = 0.252 \text{ km}^2$$

The 1974 image area for the intact rainforest was 3311 mm²:

$$X \text{ km}^2 = \frac{3311 \text{ mm}^2_{image}}{1} \times \frac{74{,}891{,}716 \text{ mm}^2}{1 \text{ mm}^2_{image}} \times \frac{1 \text{ m}^2}{1{,}000{,}000 \text{ mm}^2} \times \frac{1 \text{ km}^2}{1{,}000{,}000 \text{ m}^2} = 0.248 \text{ km}^2$$

Repeat these same computation procedures for 1987 and 2001 to determine how much change occurred in the damaged and undamaged rainforest habitats. The total ground area for all years remained 0.5 km².

5. The 1987 (Appendix E, Figure E.5) eruption-damaged rainforest image area was 42,400 mm², so its ground area was:

   = _____ km²

6. The 1987 (Appendix E, Figure E.5) intact rainforest image area was 50,096 mm², so its ground area was:

   = _____ km²

7. Absolute change of intact rainforest area from 1974 to 1987 was:

   (1987 area _____ km²) − (1974 area of 0.248 km²)   = Change _____ km²

8. The 2001 (Appendix E, Figure E.6) eruption-damaged rainforest image area was 37,923 mm², so its ground area was:

= _____ km²

9. The 2001 (Appendix E, Figure E.6) intact rainforest image area was 54,573 mm², so its ground area was:

= _____ km²

10. Absolute change of intact rainforest area from 1987 to 2001 was:

    (2001 area _____ km²) − (1987 area _____ km²)   = Change _____ km²

11. Suppose that a single nene (Hawaiian goose, *Branta sandwicensis*) requires either 0.08 km² of damaged rainforest or 0.15 km² of intact rainforest to survive. How many of these rare (~2000 left wild in 2003) geese could the 0.5 km² study area support in 1974, in 1987, and then in 2001?

	1974	**1987**	**2001**
a. Copy the km² of damaged rainforest from questions 5 and 8.	0.252 km²		
b. Copy the km² of damaged rainforest from questions 6 and 9.	0.248 km²		
c. How many geese could the damaged rainforest area support (carrying capacity)?			
d. How many geese could the intact rainforest area support (carrying capacity)?			
e. How many geese could this 0.5 km² study area support (carrying capacity) in total?			
f. If the 0.5 km² study area actually held its carrying capacity of geese, what percent of the world's population would these represent?			

12. Suppose that a single 'alala (Hawaiian crow, *Corvus hawaiiensis*) requires either 0.03 km² of damaged rainforest or 0.02 km² of intact rainforest to survive. How many of these severely endangered (~15 still alive, and only three left wild, as of 2003) crows could the 0.5 km² study area have supported in 1974, in 1987, and then in 2001?

	1974	**1987**	**2001**
a. Copy the km² of damaged rainforest from questions 5 and 8.	0.252 km²		
b. Copy the km² of damaged rainforest from questions 6 and 9.	0.248 km²		
c. How many crows could the damaged rainforest area support (carrying capacity)?			
d. How many crows could the intact rainforest area support (carrying capacity)?			

e. How many crows could this 0.5 km² study
area support (carrying capacity) in total?   _____   _____   _____

f. If the 0.5 km² study area actually held its
carrying capacity of crows, what percent
of the world's population would these represent?   _____   _____   _____

13. The U.S. National Park Service, the State of Hawaii, and numerous private and corporate sponsors have dedicated much effort to restoring nene goose and Hawaiian crow populations. The following inquiries, based upon your quantitative data, solicit your qualitative interpretations, which in the end become value judgments lacking absolute "rights or wrongs." Such decisions are now typical in modern environmental management.

   a. What was the percentage change of total carrying capacity within our 0.5 km² Devastation Trail study area for each bird between 1974 and 2001?

   Goose _____%   Crow _____%

   b. Which bird appears to fare better in intact forest than in damaged rainforest?

   c. Which bird appears to fare better in damaged forest than in intact rainforest?

   d. If necessary, which bird (goose or crow) do you think should receive conservation and habitat maintenance priority within the 0.5 km² Devastation Trail study area?

   e. Explain the rationale that underlies your habitat management decision in question 13(d).

   f. What habitat management strategy would you suggest to benefit the bird that you chose in question 13(d)?

EXERCISE 17 † IGNEOUS LANDFORMS

PURPOSE

The purpose of this exercise is to learn about igneous processes and landforms, and the tectonic settings in which these processes and landforms occur.

LEARNING OBJECTIVES

By the end of this exercise you should be able to

- describe the characteristics of different types of volcanoes, such as their morphology, eruptive behavior, and the igneous rocks associated with them;
- measure the morphological characteristics (e.g., relief, slope) of different types of volcanoes, using topographic maps;
- explain how igneous rock composition is related to the eruptive behavior of volcanoes;
- list the types of volcanoes and igneous rocks associated with different tectonic settings;
- recognize different types of igneous landforms on topographic maps and air photos; and
- explain how igneous landscapes change over time due to both igneous activity and erosional processes.

INTRODUCTION

Volcanic activity occurs primarily in association with plate tectonic activity. As tectonic plates move, molten material from the earth's interior (magma) moves up toward the earth's surface, cooling as it moves away from the interior. Molten material breaching the earth's surface results in the formation of extrusive or **volcanic igneous landforms.** These volcanic landforms may consist of lava that has cooled and solidified or of layers of ash and other airborne solid material, such as tephra, ejected from a volcano. Some molten material may never reach the surface of the earth. It may cool and solidify below the surface, forming intrusive or **plutonic igneous landforms.** At a later date, these plutons may be exposed, due to erosion of overlying material or by tectonic uplift.

A **volcano** is both the vent from which lava, pyroclasts, and gas from the earth's interior erupt at the surface, and the physical structure built up around the vent. **Lava flows** are generally composed of basaltic lava, which has a low viscosity and flows readily. **Pyroclasts**, or **tephra**, are solid chunks of material explosively ejected from a volcano and generally form from more viscous lava such as rhyolitic lava. Pyroclasts may range in size from very fine ash to large blocks the size of a house. The volcanic landforms we find, the types and composition of material ejected from a volcano, and the explosiveness of volcanic eruptions depend on the tectonic setting and the type of volcano.

Volcano Types

There are three main types of volcanoes: shield volcanoes, composite (strato) volcanoes, and cinder cones. They are distinguished on the basis of their size and shape, (Figure 17.1), the composition of the magma and associated igneous rocks, the eruptive behavior, and the landforms found in conjunction with them (Table 17.1).

Shield volcanoes are the largest of all volcanoes (Figure 17.1) and occur most often in association with hot spots located in the interior of oceanic tectonic plates (e.g., Hawaii) and on oceanic rift zones (e.g., Iceland). They have gently sloping sides and produce magma with a low silica

FIGURE 17.1 SIZE COMPARISON OF SHIELD AND COMPOSITE VOLCANOES

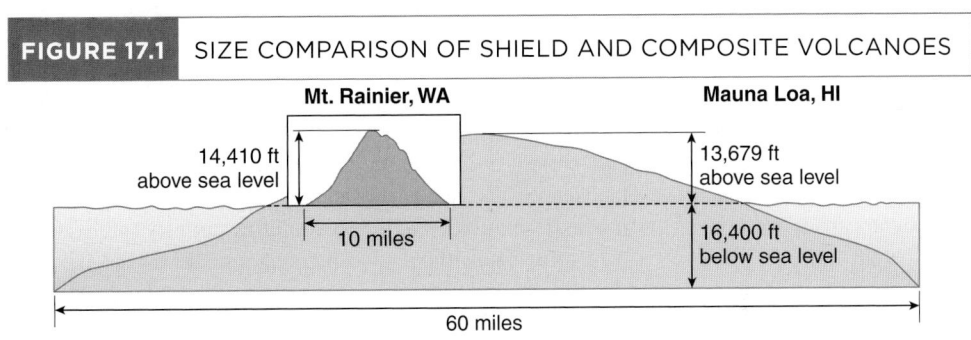

TABLE 17.1 COMPARISON OF VOLCANO TYPES

Volcano Type	Morphology: Shape and Local Relief	Tectonic Setting	Relative Silica Content and Viscosity	Eruptive Behavior	Extrusive Rock
Shield volcano	Broad, gently sloping (2–10°); Up to 10 km (6.2 mi)	Oceanic hot spots	Low silica and low viscosity	Effusive	Basalt
Strato- or Composite volcano	Moderately steep (up to 30°) symmetrical cone; Up to about 4 km (2.5 mi)	Subduction zones; Continental rift zones; Continental hot spots	Moderate to high silica and moderate to high viscosity	Primarily explosive	Andesite Dacite Rhyolite
Cinder cone	Very steep (up to 33°) linear cone; Up to about 300 m (1000 ft)	Subduction zones; Hot spots; Continental rift zones; Oceanic rift zones	Silica and viscosity vary	Intermediate	Mainly basaltic rocks

TABLE 17.2 IGNEOUS ROCKS AND ERUPTIVE BEHAVIOR

Rock:	Basalt	Andesite	Dacite	Rhyolite
Si – content:	Low	Moderate	High	Very high
Eruptive Behavior:	Effusive	Mixed effusive and explosive	Explosive	Explosive

content, mainly basalt. Because basalt is relatively low in silica, it has a low viscosity, which results in fluid-like lava and effusive (nonexplosive) eruptions (Table 17.2). Shield volcanoes are built of successive layers of lava flows, with relatively little ash. Because the viscosity of the lava is so low, the lava readily flows downhill and steep slopes do not form as the lava cools and solidifies.

Composite volcanoes, or **stratovolcanoes**, occur primarily in conjunction with subduction zones (e.g., the Cascade Range volcanoes) but may also occur in continental rift zones (e.g., Mount Kilimanjaro in Africa). These volcanoes form steep-sided, cone-shaped mountains, most of which attain sufficient height to have snow- and ice-covered peaks. They are formed of alternating layers of lava and pyroclasts. The silica content of magma from composite volcanoes ranges from intermediate to high, resulting in occasional explosive eruptions of pyroclasts, which can be quite dangerous. Lava from these volcanoes solidifies to form andesite, dacite, and rhyolite, in order of increasing silica content (Table 17.2). Due to the relatively high viscosity of this lava, it does not flow downhill readily as does lava associated with shield volcanoes, and as a result, steeper slopes form. Any snow and ice cover may result in an additional hazard during eruptions: as the snow and ice melts, it mixes with pyroclasts, soils, vegetation, and other material, forming debris flows which are also quite dangerous.

Cinder cones are also steep sided, but they are much smaller than composite volcanoes. They are made of layers of tephra and ash forcibly ejected from a vent under pressure. As these cinders fall back to earth, they pile up, forming a steep-sided hill. Basaltic lava may flow out beneath the tephra, forming lava flows extending away from the cone. Cinder cones may form in any tectonic setting that produces volcanic activity and often occur in association with larger volcanoes. The majority of material associated with cinder cones is basaltic in composition.

Volcanic Landscape Alteration

Almost all volcanoes have a **crater** at their summit, generally in the form of a circular depression, that marks the location of the vent from which material erupts. Some composite volcanoes result in such massive explosions, however, that a caldera may form. A **caldera** is a larger, circular *collapse* structure that results when enormous amounts of magma are blown out from beneath a volcano. The removal of this magma removes support for the overlying rock layers that form the sides and top of the volcano. Subsequently, the sides and top of the volcano collapse into the void once filled with magma, creating a large depression where a mountain once existed. Thus a caldera is a distinct type of crater. Crater Lake, Oregon, is a caldera, and the mountain that once existed there is now called Mount Mazama.

Ash fall can also alter the landscape by accumulating in low spots and reducing the overall relief of the landscape. Lava flows typically do the same.

Once volcanic activity stops, the processes of weathering and erosion become the dominant processes affecting the landscape. The relative resistance of igneous rocks to these processes compared to the surrounding country (preexisting) rock affects how the landscape changes over time. Likewise, the climate and topography of an area affect the degree and speed of weathering and erosion.

A **volcanic neck,** a massive tower of resistant igneous rock, may be exposed after millions of years of weathering and erosion. The volcanic neck forms in what was once the main conduit magma traveled through from the earth's interior to the vent at the surface. After cessation of volcanic activity, erosion of the outlying layers of lava and tephra comprising the volcano reveals the old conduit, composed of coarse-textured igneous rock. A volcanic neck is an example of a plutonic landform.

Likewise, dikes and sills may become visible after millions of years of weathering and erosion. **Dikes** are thin, vertical layers of magma injected into fractures that cut across layers of preexisting rock. **Sills** are thin, horizontal layers formed when magma forces its way between horizontal layers of preexisting rock. Dikes and sills may form below the surface during active volcanism with any type of volcano; but these are classified as plutonic landforms by having formed underground. After volcanic activity ceases, dikes and sills may be exposed as overlying material is eroded away. In the case of Ship Rock, New Mexico, dikes can be seen radiating outward from a volcanic neck. Figure 17.2 illustrates these igneous landforms.

IMPORTANT TERMS, PHRASES, AND CONCEPTS

volcanic landforms	cinder cone
plutonic landforms	crater
volcano	caldera
lava flow	volcanic neck
pyroclasts, tephra	shield volcano
dike	sill
composite or stratovolcano	

FIGURE 17.2 INTRUSIVE IGNEOUS LANDFORMS

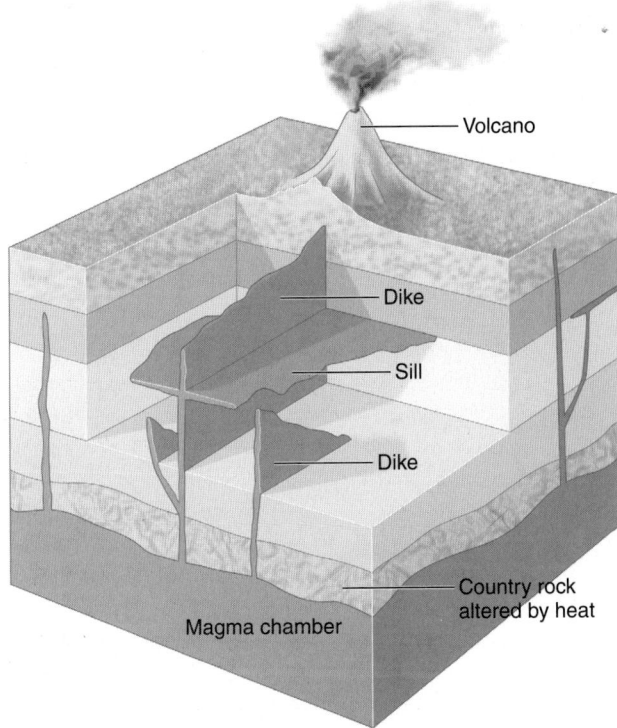
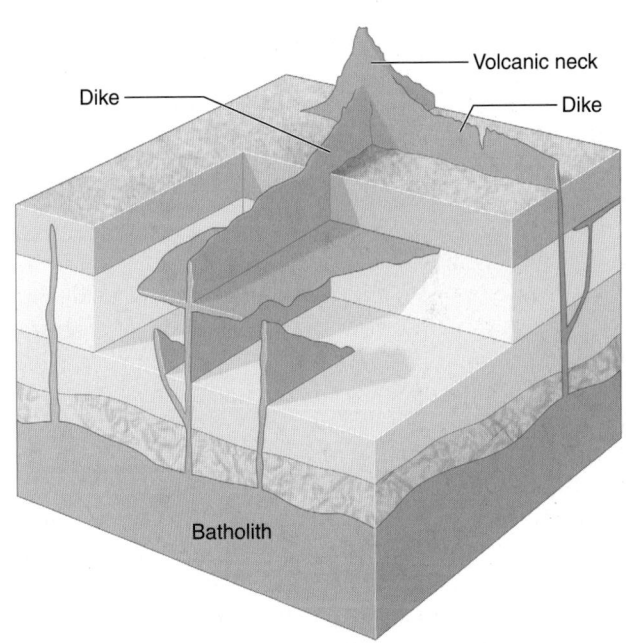

Source: Adapted from Lutgens et al., 2000.

Name: _____ Section: _____

IGNEOUS LANDFORMS

1. Explain briefly how a caldera forms and how it is different from other types of volcanic craters.

2. Use the topographic map of Crater Lake (Appendix E, Figure E.11) to determine the approximate diameter of Crater Lake in miles. Measure from the shoreline by The Watchman on the west side of the lake to the shoreline by Skell Head on the east side of the lake. Use the representative fraction to calculate the diameter. Please show all your work.

3. Determine the total depth of the crater that contains Crater Lake (i.e., the distance from the *top* of the crater rim to the *bottom* of the lake) by working through parts (a), (b), and (c).

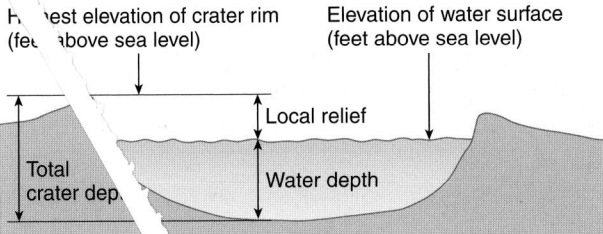

   a. Approximately how deep is the water in Crater Lake, in feet? _____

   b. Determine the local relief from the water level in Crater Lake to the top of the crater rim, in feet. Examine the contour lines carefully to determine where the ridge defining the rim of the crater is located. Then look for the benchmark designating the highest point along the crater rim.

   What is the highest benchmark elevation along the crater rim? _____

   What is the elevation of the water surface? _____

   What is the local relief from the water surface to the top of the crater? _____

   c. Using the answers to parts (a) and (b), what is the total depth of the crater, in feet?

4. Use the topographic map of Mount St. Helens (Appendix E, Figure E.12) to measure

   a. the diameter of the crater in miles, using the representative fraction. Measure this from the red "x" on the west crater rim, through the center of the lava dome, to the red "x" on the east crater rim. Show your work.

209

b. the total depth of the crater from the highest point on the crater rim (look for a benchmark) to the bottom of the crater (look at the contour lines), in feet.

5. Use the topographic map of Capulin Mountain (Folsom, New Mexico, Appendix E, Figure E.13) to measure

   a. the diameter of the crater in miles, using the representative fraction. Measure this from the red "x" on the west crater rim, through the center of the crater, to the red "x" on the east crater rim. Show your work.

   b. the total depth of the crater from the highest point on the crater rim (look for a benchmark) to the bottom of the crater (look at the contour lines), in feet.

6. Determine the slope angle for Capulin Mountain (Appendix E, Figure E.13) by following these steps.

   a. Select an index contour line that approximates the base of the northeastern side of the mountain and determine its elevation.

   Approximate base elevation (feet) _____

   b. The highest elevation of the crater rim is a benchmark with an elevation of 8182 feet. What is the local relief from here to your answer to part (a)?

   Local relief (feet) _____

   c. Draw a straight line starting at this benchmark on the crater rim, going toward the northeast, and ending at the contour line you selected to approximate the base of the mountain. Calculate the length of this line in feet, using the representative fraction. Show your work.

   Distance from rim to base (feet) _____

   d. Use your answers to parts (b) and (c) to calculate the gradient in feet/foot.

   $$\text{Gradient} = \frac{\Delta \text{ Elevation}}{\text{Distance}} = \underline{\hspace{2cm}}$$

   e. The inverse tangent of the gradient in feet/foot (the answer to part d) will tell you the slope in degrees. If your calculator has trigonometric functions, this is the tangent^{-1} function, but make sure your calculator is set to degree mode before doing this. If your calculator doesn't have trigonometric functions, see Appendix A for a table of tangent values and their associated gradients in degrees.

   Slope in degrees = _____

7. Determine the slope angle for Mauna Loa, Hawaii (Appendix E, Figure E.14). Follow the same procedure outlined in question 6, measuring from the highest benchmark on the crater rim due west to the edge of the island.

   Highest elevation (ft) _____

   Base elevation (ft) _____

   Local relief (ft) _____

   Distance (ft) _____

   Gradient (feet/foot) _____

   Slope (degrees) _____

8. In the spaces that follow, summarize your measurements of crater width and depth, local relief, and slope. Together, these measurements describe the morphology of these volcanoes. Then use the information in Table 17.1 describing the morphology of the various types of volcanoes to determine what type of volcano each of these is — shield, composite, or cinder cone.

	Crater Width (mi)	Crater Depth (ft)	Local Relief (ft)	Slope (degrees)	Volcano Type
Crater Lake			2151	12°	
Wizard Island	0.06	90	762	27°	
Mount St. Helens			4765	21°	
Mount Rainier	0.3	270	9410	19°	
Capulin Mountain					
Mauna Loa	2.4	600			

9. Crater Lake (formerly Mount Mazama), Wizard Island, Mount St. Helens, and Mount Rainier are all part of the Cascade Range volcanoes, and thus are all a result of the same tectonic setting. Mount Rainier is the most "intact" of these volcanoes; its morphology is more typical of a volcano that has not been altered by extreme recent volcanic activity.

   a. Figure 17.3 shows the tectonic setting of the Cascade Mountain Range. What type of plate boundary is causing volcanic activity in this region?

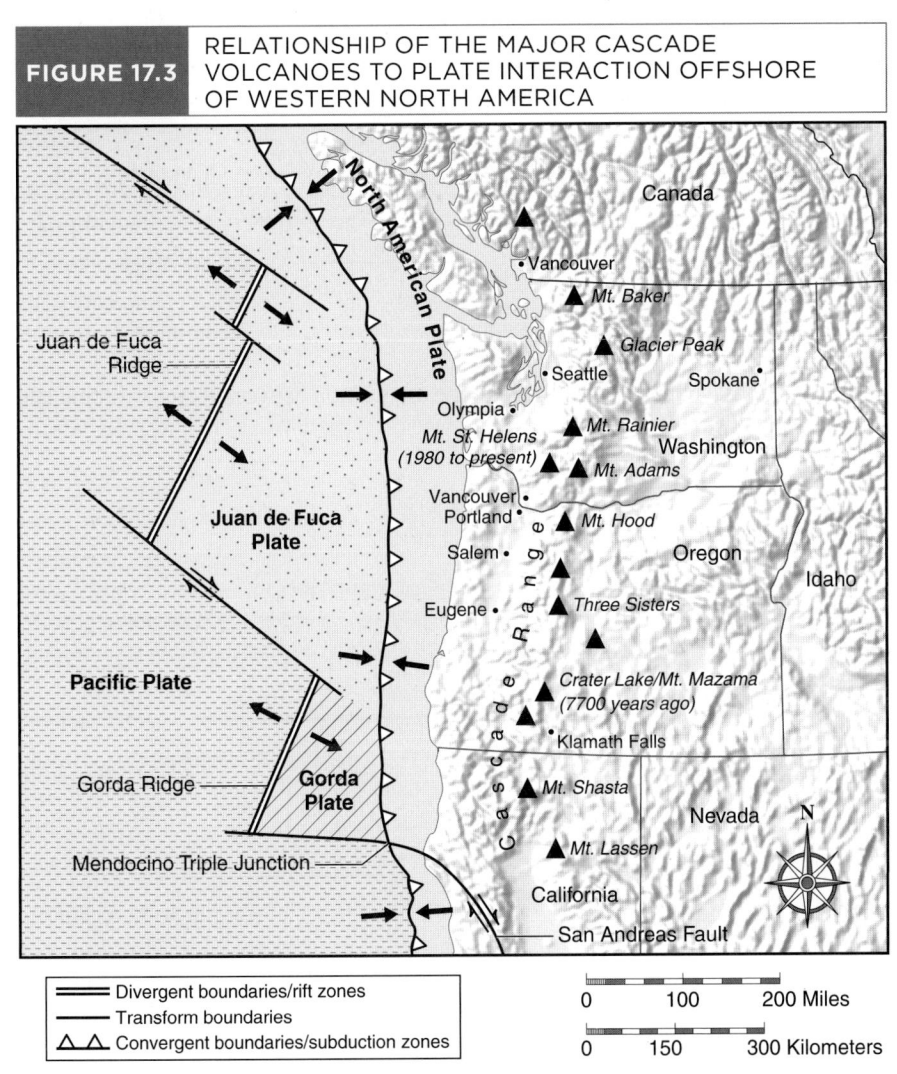

FIGURE 17.3 RELATIONSHIP OF THE MAJOR CASCADE VOLCANOES TO PLATE INTERACTION OFFSHORE OF WESTERN NORTH AMERICA

b. Based on your answers to question 8, what types of volcanoes are associated with this type of tectonic setting?

c. Comparing the size of the craters of Crater Lake, Wizard Island, Mount St. Helens, and Mount Rainier, do you think the crater at Mount St. Helens is most likely an example of a caldera or is it just a large crater? Why?

10. The kind of igneous rocks found in association with volcanoes varies depending on the type of volcano and the tectonic setting. Use the topographic and geologic maps listed, and for each landform listed, determine what type of volcanic landform it is and the dominant type of volcanic rock or rocks associated with that particular landform.

	Type of Volcanic Landform	Igneous Rock(s)	Tectonic Setting
Wizard Island (Figures E.11 and 17.4)	_____	_____	_____
Red Cone (Figure E.11 and 17.4)	_____	_____	_____
Crater Lake (Figure E.11 and 17.4; look for the category that covers the greatest land area adjacent to Crater Lake)	_____	_____	_____
Mount St. Helens (Figures E.12 and 17.5)	_____	_____	_____
Mauna Loa (Figure E.14)	_____	Basalt	Oceanic hot spot
Capulin Mountain (Figures E.13 and 17.6)	_____	_____	Ancient continental rift zone
Ship Rock (Figures E.15 and 17.7)	_____	_____	Ancient continental rift zone

11. Based on your answers to question 10, what correlation appears with regard to the association between tectonic setting and igneous rock type, and volcanic landform and igneous rock type?

Tectonic Setting	Igneous Rock(s)
Subduction zone	_____
Oceanic hot spot	_____
Continental rift zone	_____

Igneous Landform	Igneous Rock(s)
Cinder cone	_____
Caldera	_____
Composite volcano	_____
Shield volcano	_____

12. Based on the rock types you listed in question 11, would you expect eruptions in these tectonic settings to be primarily effusive, explosive, or a mix of the two?

subduction zones _____

continental rift zones _____

oceanic hot spots _____

FIGURE 17.4 GENERALIZED GEOLOGIC MAP OF CRATER LAKE AND VICINITY

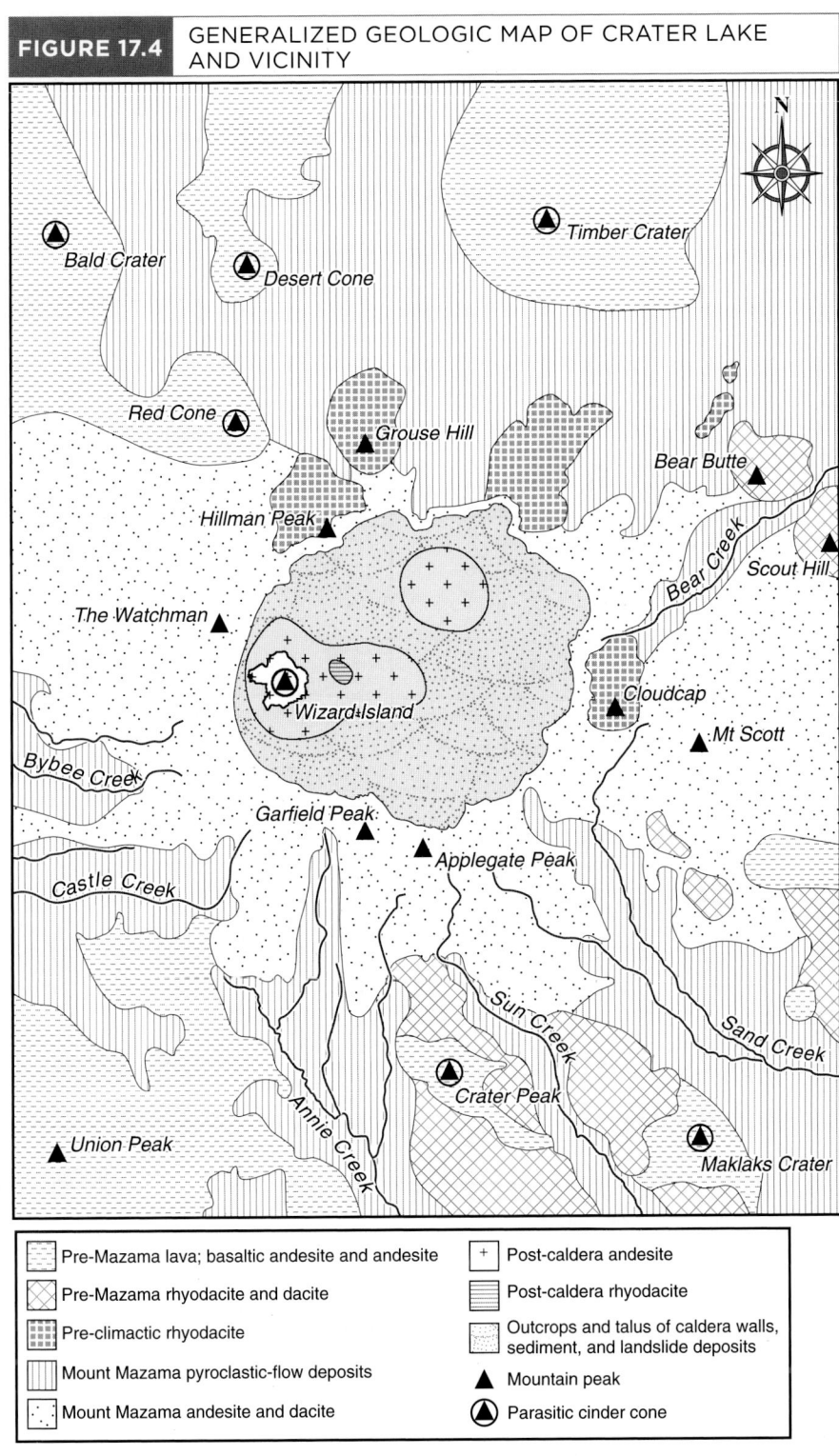

Pre-Mazama lava; basaltic andesite and andesite	+ Post-caldera andesite
Pre-Mazama rhyodacite and dacite	Post-caldera rhyodacite
Pre-climactic rhyodacite	Outcrops and talus of caldera walls, sediment, and landslide deposits
Mount Mazama pyroclastic-flow deposits	▲ Mountain peak
Mount Mazama andesite and dacite	Parasitic cinder cone

Source: After Howell 1942, Topinka 2001, and USGS/CVO.

FIGURE 17.5 GENERALIZED GEOLOGIC MAP OF MOUNT SAINT HELENS' BLAST ZONE

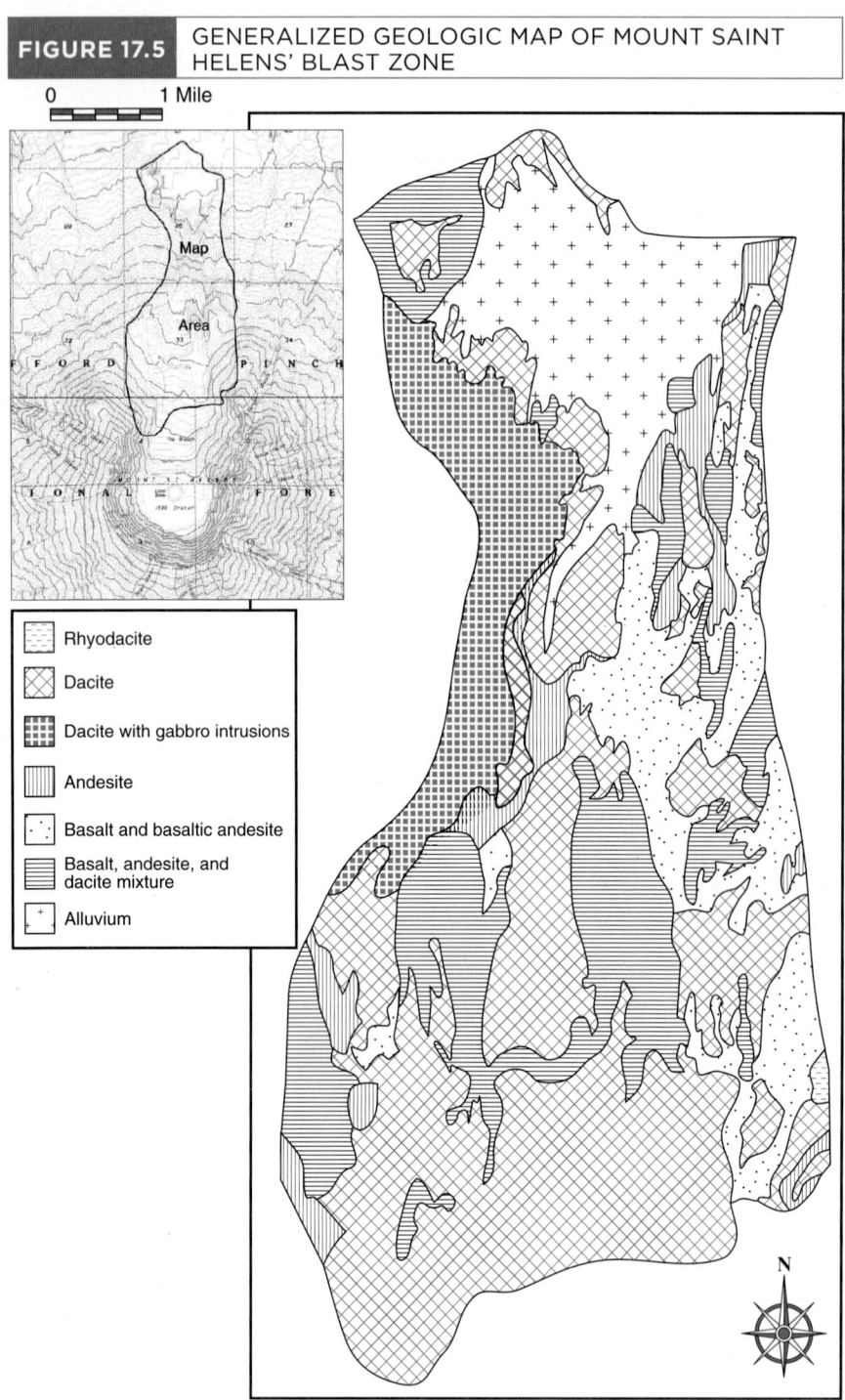

Source: After Hausback 2000, USGS.

13. According to Figure 17.6, Capulin volcano is composed of basalt, and basalt is the dominant rock type in this rift zone; however, andesite and dacite are also present. If we mapped igneous rocks along oceanic rift zones, we would find only basalt. What does this suggest about eruptive behavior along oceanic rift zones compared to continental rift zones?

14. Examine the geologic map of Crater Lake (Figure 17.4) and the key describing the various deposits.

    a. According to the key, which two geologic units are the oldest?

FIGURE 17.6 — GENERALIZED GEOLOGIC MAP OF PART OF THE RATON-CLAYTON VOLCANIC FIELD

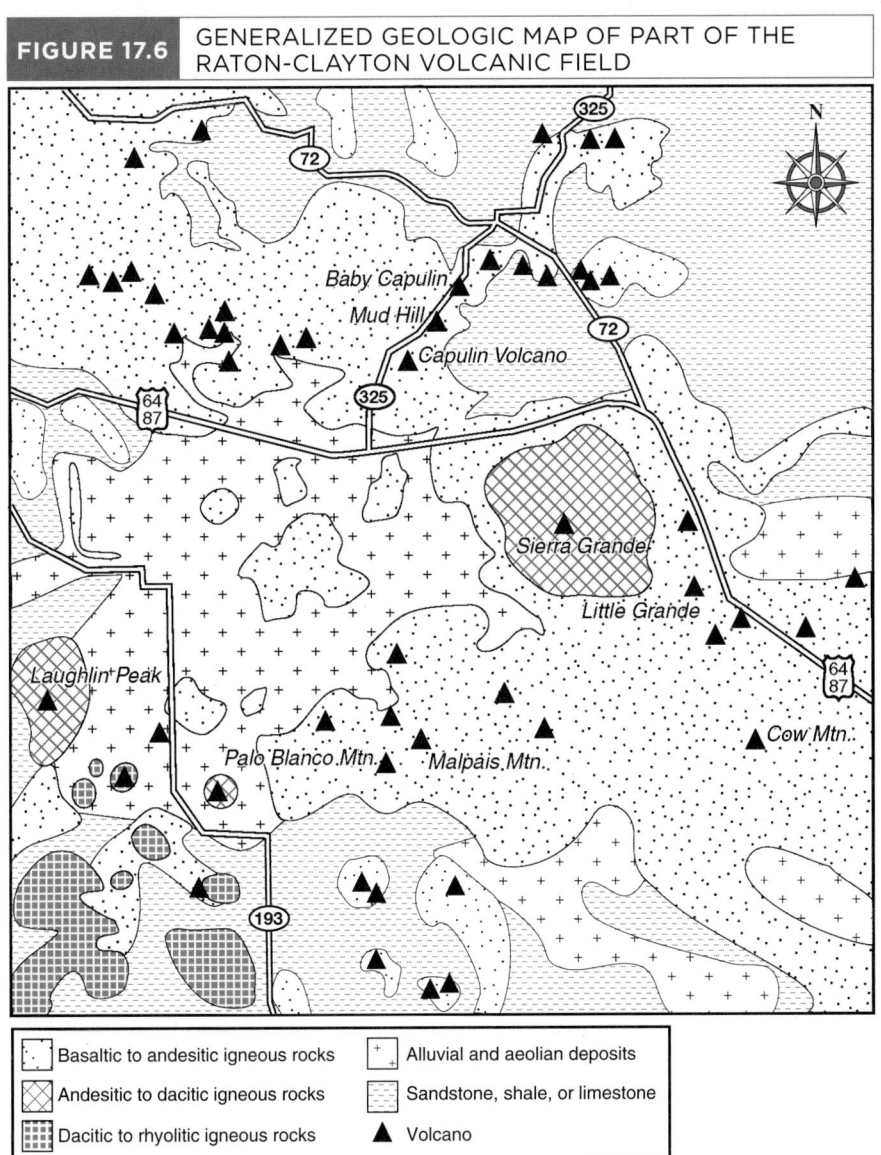

Source: After geologic map of New Mexico, New Mexico Bureau of Geology and Mineral Resources, 2003; and USGS.

b. Note that these two oldest units are not located adjacent to Crater Lake; they are located farthest from the lake. Presumably these old deposits were also located nearer to Crater Lake in the past than they are today. Why do you think these old deposits are no longer found adjacent to Crater Lake?

c. What named landform was created after Mount Mazama erupted and Crater Lake had already formed? How do you know this?

d. What processes do you think are modifying the landscape today?

215

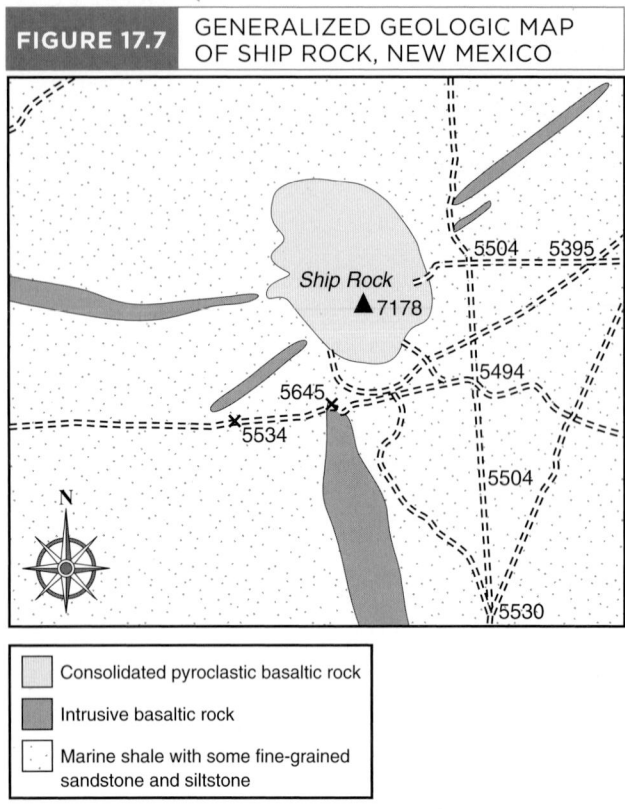

FIGURE 17.7 GENERALIZED GEOLOGIC MAP OF SHIP ROCK, NEW MEXICO

- Consolidated pyroclastic basaltic rock
- Intrusive basaltic rock
- Marine shale with some fine-grained sandstone and siltstone

Source: After O'Sullivan et al., 1963, USGS.

15. Examine the map of ash deposits associated with Crater Lake (Figure 17.8).

   a. Based on the isolines showing ash thickness, how thick are the thickest deposits of ash, and where do we find these thickest deposits in relation to Crater Lake (N, S, E, W)?

   b. Based on the isolines showing the ash thickness, what was the prevailing wind direction at the time Mount Mazama erupted? How did you determine this?

   c. Examine the topography to the south of Crater Lake, and the topography to the north and northeast (Appendix E, Figure E.11). Which area is flatter, the area to the north and northeast or the area to the south?

   d. What might be the reason for this difference in topography? Use Figures 17.4 (Geologic Map) and 17.8 (Ash Map) to help answer this question, and think about what the topography to the north and to the south of Crater Lake might have been like prior to the eruption that created the lake.

FIGURE 17.8 DISTRIBUTION OF ASH FALL AND PYROCLASTIC FLOW DEPOSITS FROM THE ERUPTION OF MOUNT MAZAMA

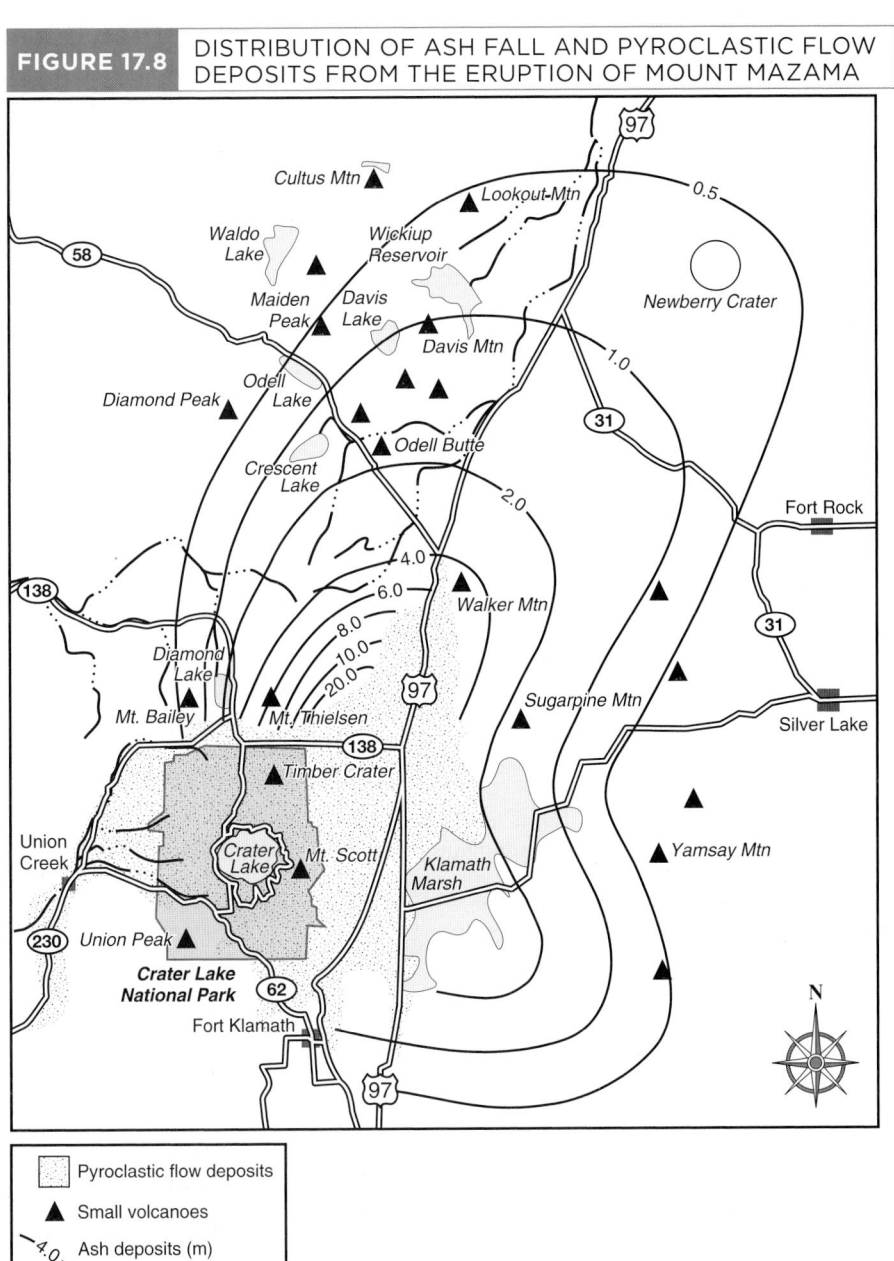

Source: After Sherrod et al., 2000, USGS.

16. Figure 17.9 shows the ash fallout from the 1980 eruption of Mount St. Helens.

   a. Based on the patterns shown on the map, what was the prevailing wind direction at the time of the eruption? How did you determine this?

   b. How does the thickness of ash fall from Mount St. Helens compare to the thickness of ash fall from Mount Mazama?

FIGURE 17.9 DISTRIBUTION OF ASH FALL FROM MOUNT ST. HELENS' 1980 ERUPTION

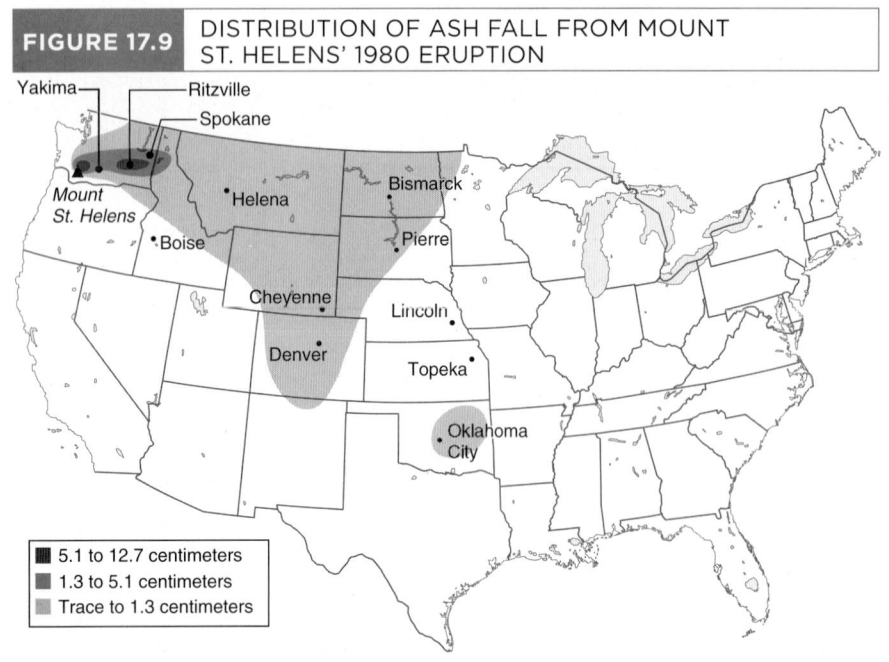

- 5.1 to 12.7 centimeters
- 1.3 to 5.1 centimeters
- Trace to 1.3 centimeters

Source: Adapted from Topinka, 1997, USGS/CVO.

    c. Examine the stereo photos of Mount St. Helens (Figure 17.10). Refer to Appendix D for information on using pocket stereoscopes to view stereo photos in three dimensions. Describe the topography to the north of the blast zone (north of the light grey area).

    d. What do you think the relief of this landscape to the north of Mount St. Helens would look like had the ash fall from Mount St. Helens been similar to that of Mount Mazama?

17. a. Examine the topographic map of Ship Rock, New Mexico (Appendix E, Figure E.15). Note the two long ridges extending away from Ship Rock. Highlight these two ridges.

    b. What type of landform are these two ridges examples of?

    c. Examine the stereo-pair of Ship Rock (Figure 17.11). Refer to Appendix D for information on using pocket stereoscopes to view stereo photos in three dimensions. How many ridges can be seen radiating outward from Ship Rock on the air photos?

FIGURE 17.10 STEREO AIR PHOTOS OF MOUNT ST. HELENS

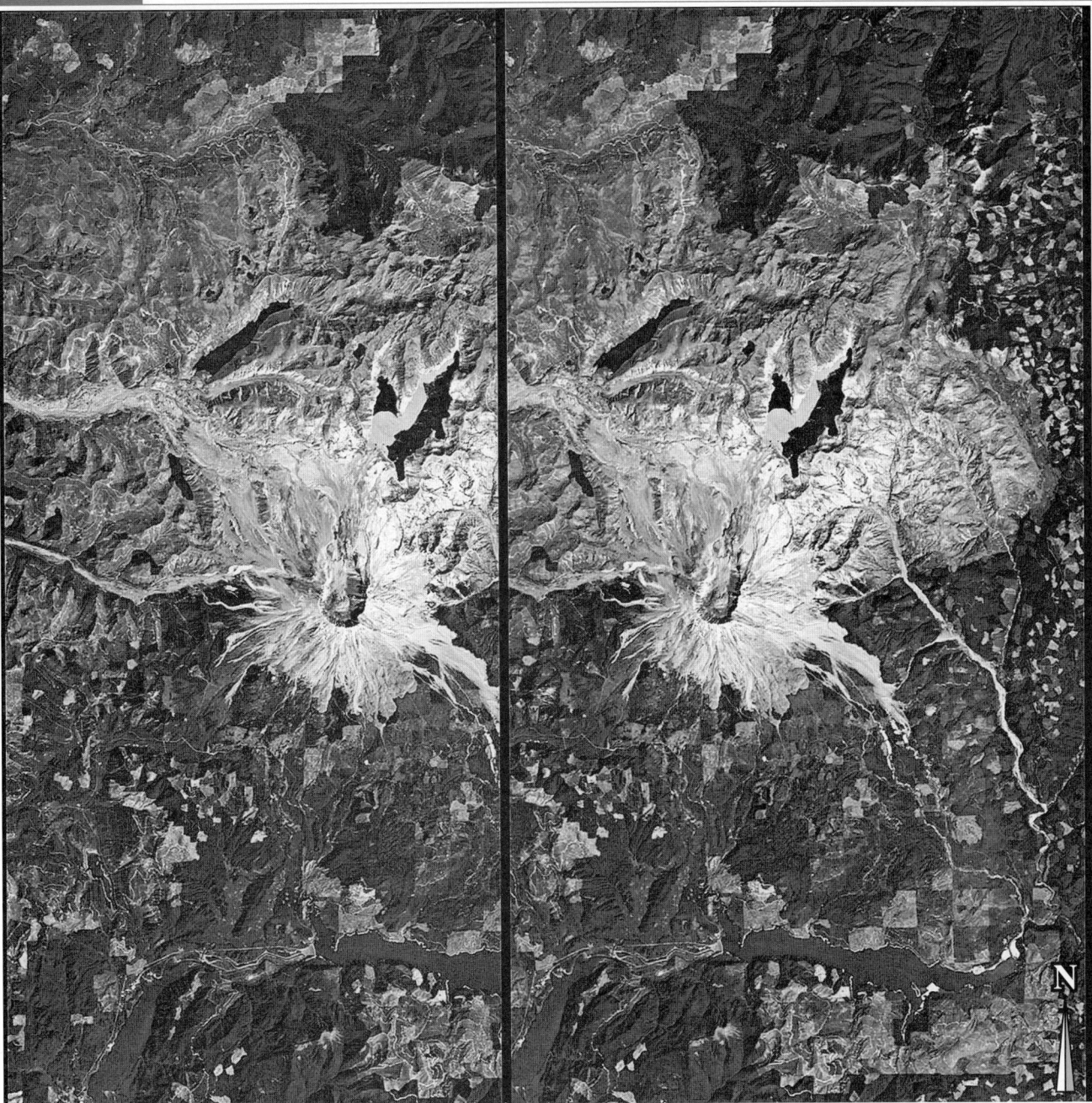

Source: Courtesy of NASA/JPL-Caltech.

d. On the Ship Rock air photos (Figure 17.11), north is to the left. Rotate your topographic map (Appendix E, Figure E.15) so that north is to the left. This way the map and the photos are oriented the same way. On your topographic map, locate and highlight the additional ridges that appear on the air photos.

e. Based on the pattern of the contour lines defining these ridges, if you walked along the top of them, would the top be smooth or irregular? How did you determine this?

FIGURE 17.11 STEREO AIR PHOTOS OF SHIP ROCK

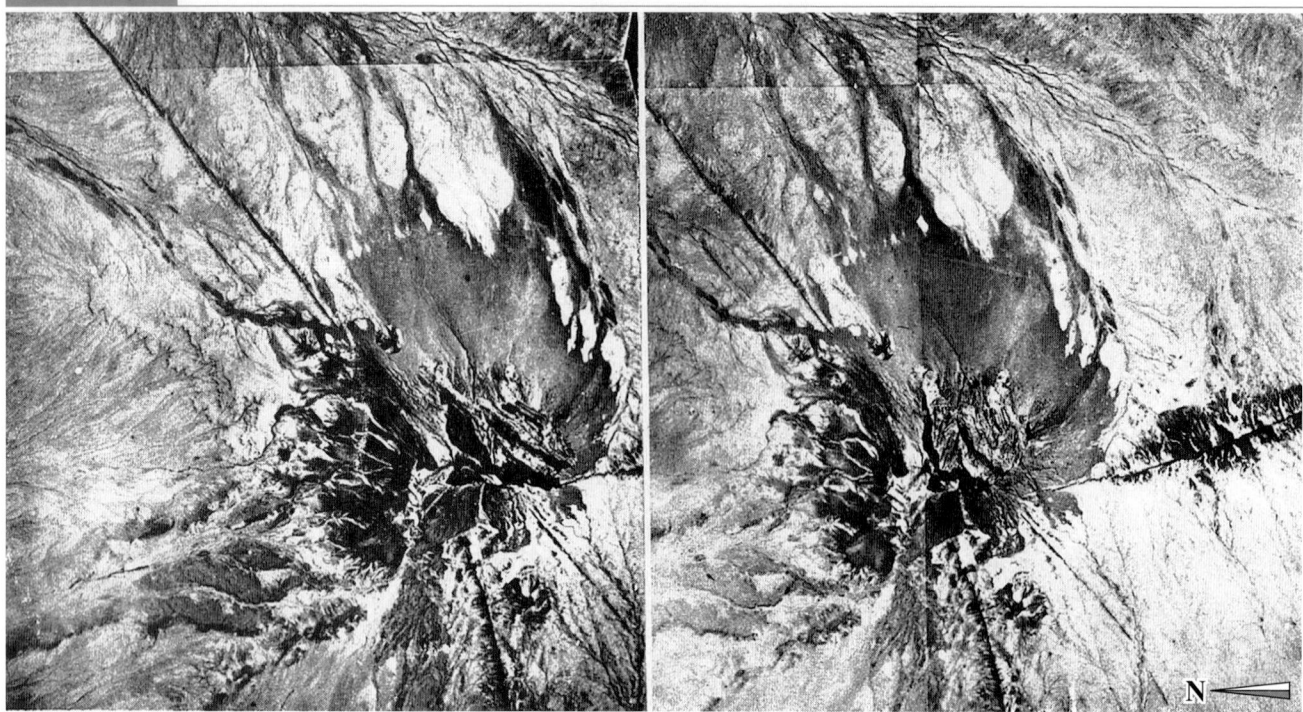

Source: Photos used with permission of American Educational Products, Ft. Collins, CO.

    f. According to the geologic map of Ship Rock (Figure 17.7), what type of rocks are these ridges composed of, and what type of rocks do we find surrounding the ridges?

        ridge rocks: _____

        surrounding rocks: _____

    g. Which of these rock types do you think is most susceptible to weathering and erosion? How did you decide this?

18. Assuming the base of Ship Rock is 5500 feet, what is the approximate local relief of Ship Rock?

19. a. What is the major process that has modified the landscape following the end of volcanic activity, producing the features evident in the area of Ship Rock today?

    b. Why has this process not erased all traces of the volcano that once existed here?

20. Based on the landscape surrounding Ship Rock today and the landscape surrounding Crater Lake today, do you think the volcanic activity at Ship Rock occurred at about the same time as, long before, or long after that at Crater Lake? Why?

EXERCISE 18 · DRAINAGE BASIN ANALYSIS

PURPOSE

The purpose of this exercise is to learn about drainage basins and their response to hydrologic inputs.

LEARNING OBJECTIVES

By the end of this exercise you should be able to

- delimit drainage divides on topographic maps;
- identify linear, concave, and convex slopes (and thus valleys and drainage divides) on topographic maps by examination of contour line shape and spacing;
- explain the impacts of drainage basin characteristics and climate on the generation of surface runoff and erosion;
- relate the shape of a hydrograph to relative amounts of infiltration, surface runoff, and throughflow, and to drainage basin characteristics such as soils, vegetation, land use, and topography;
- calculate and interpret flood frequency statistics; and
- explain the relationship between flood magnitude and flood frequency.

INTRODUCTION

Drainage basins are the fundamental unit of study in hydrology, water resources, and fluvial geomorphology. A **drainage basin** is the total land area drained by a network or system of connected streams (Figure 18.1). All precipitation entering a drainage basin exits the basin either by evapotranspiration, by becoming part of the regional groundwater flow, or by flowing through the stream network to the basin outlet. The **outlet** or **mouth** of a drainage basin is the lowest point in the stream network. All streamflow leaving the basin must pass this point. The amount of water flowing through the outlet of a drainage basin, the speed of the flow, and the variations in flow over time, are a function of all the hydrological processes operating within the basin, such as evaporation, infiltration, throughflow, and overland flow (surface runoff). Likewise, the amount and type of material carried by streams, whether in solution or suspension, are also a function of the hydrologic processes at work within the basin.

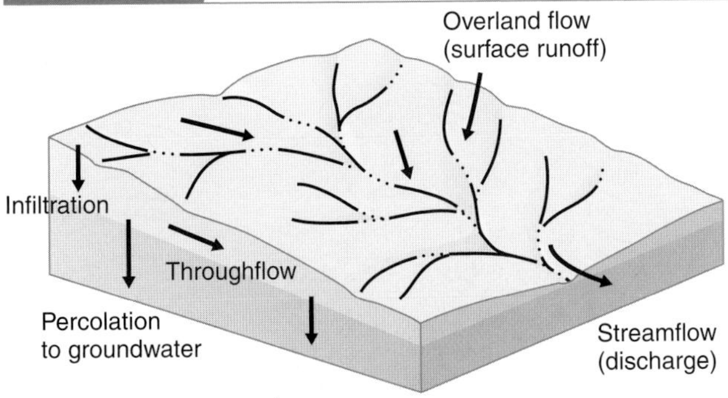

The magnitude of these hydrological processes is affected by drainage basin characteristics. For example, soil texture and permeability may affect the amount of surface runoff and subsequent increases in streamflow. Likewise, vegetation characteristics may affect the amount of interception, which, in turn, may affect the amount of surface runoff available for erosion. Thus, to understand the response of a drainage basin to storm and other events, scientists need to know the basin's characteristics.

Delimiting Drainage Basins

Topographic maps provide an excellent tool for identifying stream networks and delimiting drainage basins. The first step in delimiting a drainage basin is to identify the stream network, including intermittent streams, draining the land area of interest. The second step is to define the boundary to the basin. The boundary is defined by topographic highs, such as ridges and hill tops, shown on topographic maps by closed contour lines. Precipitation falling on one side of a ridge will flow to the stream network within the drainage basin; precipitation falling on the other side will flow to some other stream network. These high points defining the boundary to a drainage basin are called **drainage divides.** Figure 18.2 provides an example of several stream networks and the boundaries to the basins they drain.

Although many ridge tops defining drainage divides are shown on topographic maps with closed contour lines, some are not. Figure 18.3 shows some idealized hillslope shapes as block diagrams and the contour line patterns that would define these shapes. The *shape* and *spacing* of the contour lines defines the shape of the hillslope. In order to delimit

FIGURE 18.1 A SIMPLE DRAINAGE BASIN

FIGURE 18.2 | STREAM NETWORKS, DRAINAGE DIVIDES, AND DRAINAGE BASINS

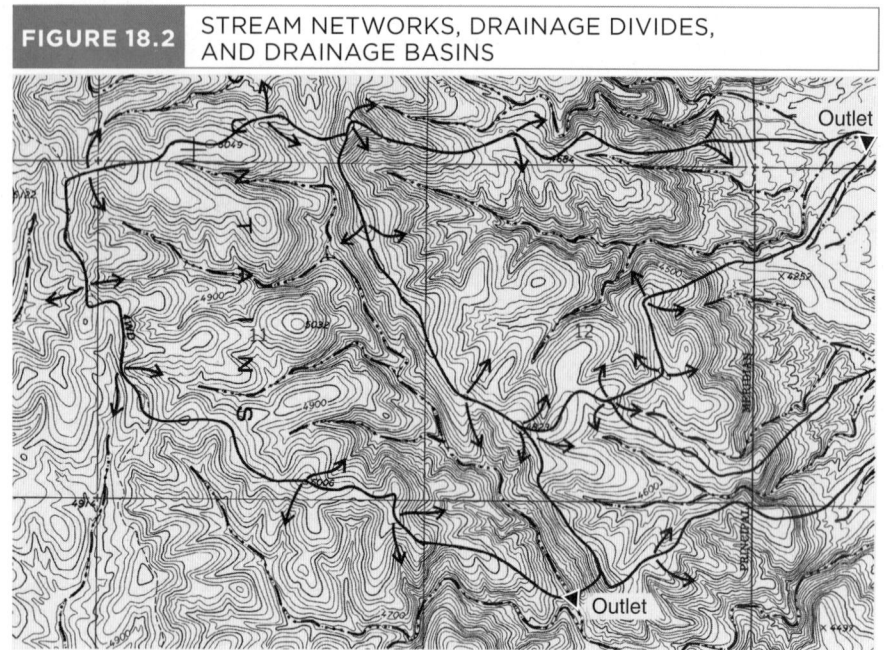

All streams are identified by a dash and three dots, drainage divides and drainage basin boundaries are identified by a solid line, and arrows show the direction of surface runoff. Triangles mark drainage basin outlets.

Source: USGS topographic map Leasburg, New Mexico 1:24,000.

FIGURE 18.3 | GEOMETRIC FORMS OF HILLSLOPES

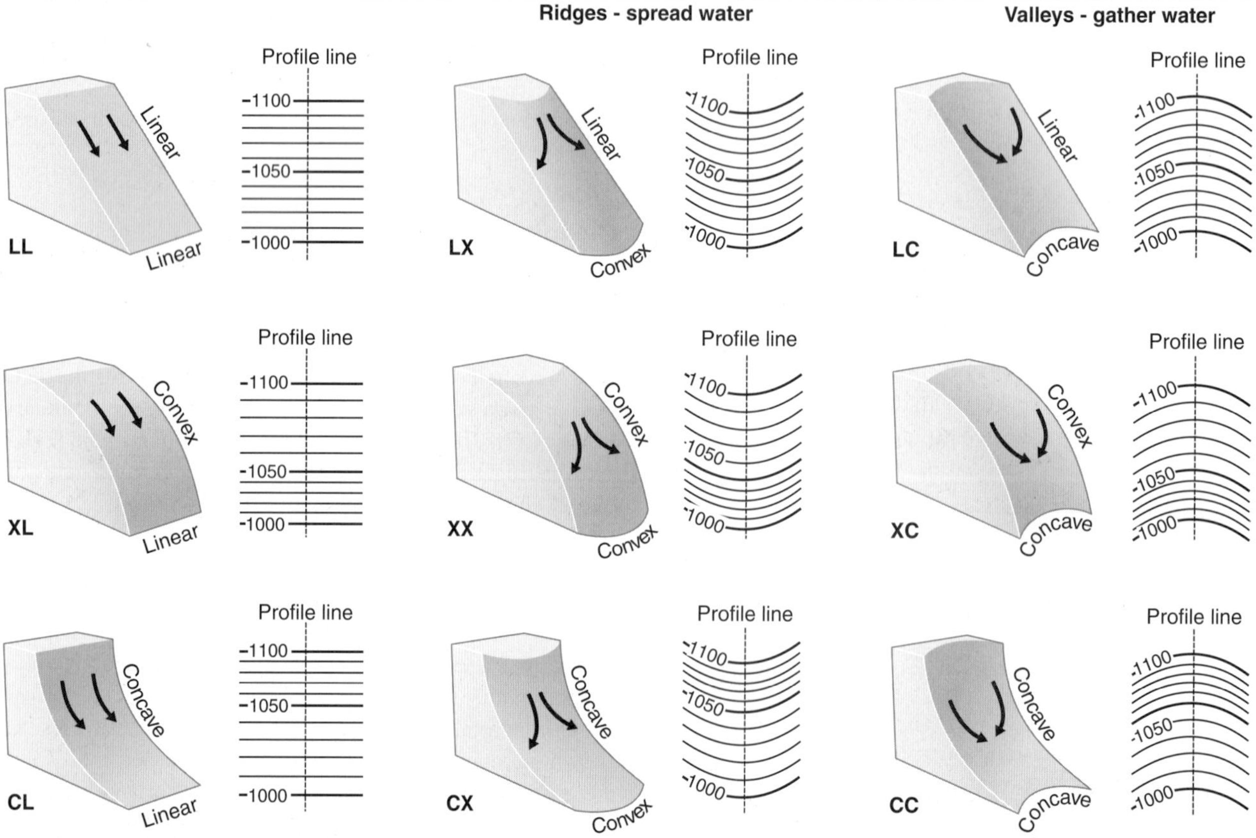

For the letter codes below each diagram, the first letter indicates the vertical shape, while the second letter indicates the horizontal shape; L = linear, C = concave, and X = convex. Arrows indicate water flow on the slopes.

Source: After NRCS, 2002.

222

drainage basins, it is necessary to be able to define all drainage divides not just those defined by closed contour lines; thus, good topographic map-reading skills are important.

A hillslope may be curved or straight in two dimensions, the vertical dimension and the horizontal dimension. Shape in the vertical dimension (from top to bottom) is shown by the *spacing* of the contour lines, and this spacing is reflected in topographic profiles. A **concave** surface is one that curves inward, and a **convex** surface is one that curves outward. The first row in Figure 18.3 shows three slopes that are *linear* in the vertical dimension; the linearity is reflected in the even *spacing* of the contour lines along the profile line. The second row in Figure 18.3 shows three slopes that are *convex* in the vertical dimension; the contour lines are farther apart at the top of the hill and closer together at the bottom. The third row in Figure 18.3 shows three slopes that are *concave* in the vertical dimension; the contour lines are closer together at the top of the hill and farther apart at the bottom. Thus, the spacing of contour lines along a profile line indicates the shape of the profile and, therefore, the shape of the hillslope from top to bottom.

Hillslope shape in the horizontal dimension (from one side of a hill to the other) is revealed by the *shape* of the contour lines as they cross the profile line. A slope that is concave (curves inward) from side to side defines a valley, while a slope that is convex (curves outward) from side to side defines a ridge. In Figure 18.3, the first column contains three slopes that are *linear* from side to side because the contour lines are straight as they cross the line defining each profile. Water will tend to flow straight down these slopes. The second column in Figure 18.3 shows three slopes that are *convex* from side to side—they are ridges—because the contour lines are bending downhill (forming V's pointing toward the bottom of the slope—just the opposite of valleys). These ridges may form drainage divides, thus water will tend to spread away from these ridges to the adjacent valley bottoms. When delimiting drainage basins, some drainage divides may appear as closed contour lines, but others may appear as ridges defined by contour lines bending downhill. The third column in Figure 18.3 shows three slopes that are *concave* from side to side—they are valleys—because the contour lines are bending uphill (toward the top of the slope). This is analogous to the rule of V's—as contour lines cross rivers and streams, they form V's bending (or pointing) upstream (uphill). Not all valleys contain rivers, but the shape of the contour lines indicates where valleys are located. These hillslopes will tend to concentrate water at the valley bottom.

The Hydrologic Cycle and Drainage Basin Characteristics

The **hydrologic cycle** describes the various paths water follows as it moves around the earth (Figure 18.4). Liquid water moving over the surface of the earth, called **overland flow** or **surface runoff,** results in increased streamflow and

FIGURE 18.4 ELEMENTS OF THE HYDROLOGIC CYCLE

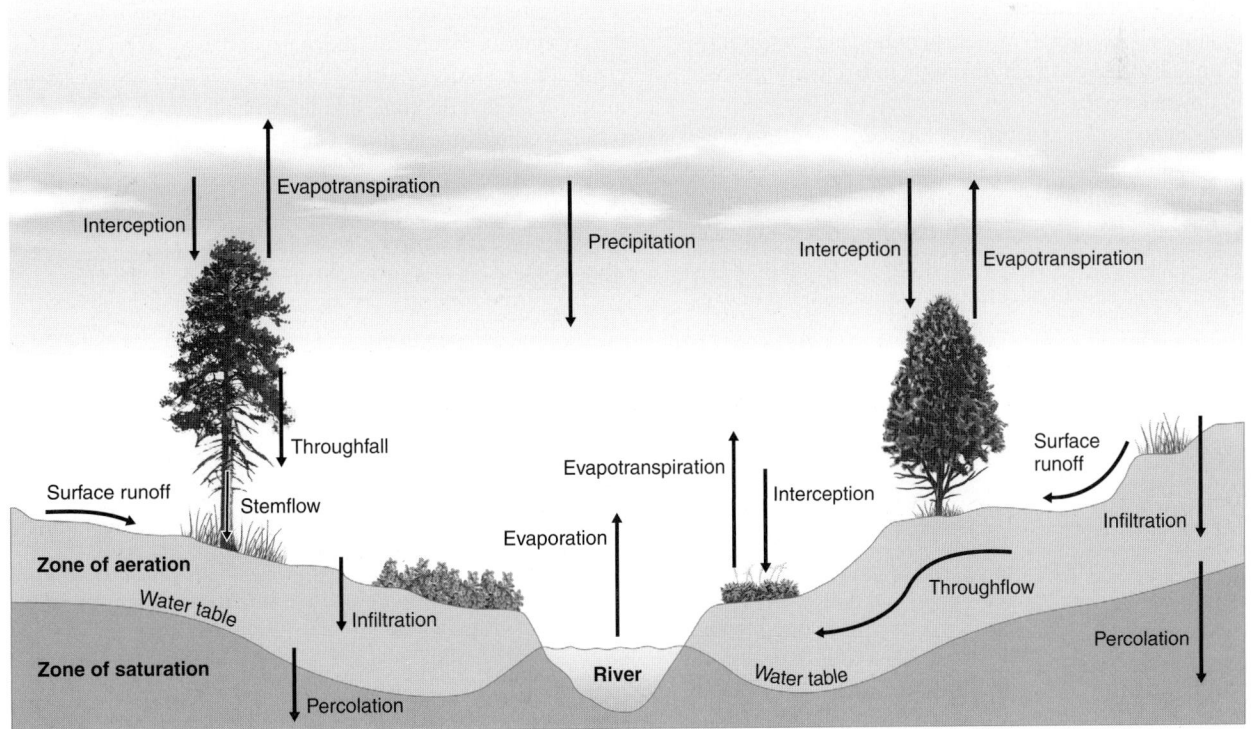

223

if sufficient in quantity, floods. Surface runoff may also alter the shape of the landscape by the processes of **erosion,** the detachment and transportation of material from one place to another, and subsequent **deposition.** Thus, storm events generate a response that involves both water and sediment movement within and out of drainage basins.

Drainage basin characteristics affecting the amount and timing of streamflow, erosion, and deposition include soils, vegetation, land use, and topography. Knowledge of these characteristics allows us to compare processes and responses of drainage basins occurring in different types of physical environments or with different characteristics.

Soil characteristics that affect the movement of water include soil texture and structure. These two characteristics affect the **permeability** of soil, the ease with which water can move through the soil, which in turn affects the **infiltration rate,** the rate at which water soaks into the soil. In order for infiltration to occur in saturated soil, the water already in the soil must percolate downward under the influence of gravity, or it must flow laterally through the soil as *throughflow*. If the rate of precipitation is greater than the infiltration rate, or if precipitation falls on already saturated soil, overland flow (surface runoff) may occur. As overland flow increases, the potential for soil erosion and for flooding both increase.

Vegetation affects the movement of water partially through the process of **interception,** the temporary storage of water on the leaves and stems of plants. Interception storage allows more time for evaporation to occur, and this reduces the total amount of water reaching the ground. The less precipitation reaching the ground, the less likely surface runoff and erosion will occur. The amount and type of vegetation in a particular region will affect the amount of interception storage (Table 18.1). In addition, plant roots help hold the soil in place; they help maintain high infiltration rates. Plants also are obstructions to the movement of water and soil downslope.

Land use and land management practices also affect rates of surface runoff and erosion. Bare soil is easily compacted, resulting in low infiltration rates and high rates of surface runoff. Without anything to hold the soil particles in place, erosion rates may also be high. Agricultural areas experience a variety of runoff and erosion rates depending on the time of year, the type of crops grown, and the conservation practices used. Urban areas have low erosion rates (no exposed soil) but high rates of surface runoff (water cannot infiltrate paved or cement-covered surfaces).

Topographic characteristics important to the production of surface runoff and of erosion are *slope gradient* and *slope length*. Steep slopes result in higher erosion rates than gentle slopes because water has less opportunity to infiltrate and because the velocity of overland flow is determined, to some extent, by slope gradient. As slope gradient decreases, the velocity of surface runoff may decrease and deposition may occur before the sediment reaches a stream channel. Slope length is important because long slopes allow more opportunity for surface runoff to concentrate in rills or gullies, which may result in extensive erosion.

Drainage Basin Response

Drainage basin response is measured at the outlet or mouth of the basin. **Stream discharge** is defined as the volume of water passing a given point per some unit of time and is represented in equations and on graphs as the letter Q. Discharge is usually measured as cubic feet per second (cfs) or cubic meters per second (cms). Stream discharge and water quality measurements, such as sediment concentration, are monitored at numerous rivers around the country by the U.S. Geological Survey. This information is analyzed to determine how drainage basins respond to storm events.

Generally, drainage basins with similar characteristics should respond in a similar fashion to storm events. Therefore, if the characteristics of a basin are known, its response to precipitation events can be estimated. Likewise, if the characteristics of the basin change over time, the change in response may be estimated. Table 18.2 lists the major factors affecting drainage basin response (surface runoff, streamflow, sediment concentration).

TABLE 18.1 PERCENT PRECIPITATION INTERCEPTED FOR SELECTED VEGETATION

Vegetation	% Interception
Grass	55
Spruce	30-45
Pine	25-30
Mixed Broadleaf	15-25

TABLE 18.2 FACTORS INFLUENCING DRAINAGE BASIN RESPONSE

Runoff characteristics	depth of overland flow, velocity of overland flow
Precipitation characteristics	amount of precipitation, intensity of precipitation
Soil characteristics	texture, structure, erodibility
Vegetation characteristics	density of vegetation cover, type of vegetation
Topographic characteristics	slope gradient, slope length
Land use characteristics	e.g., urban, rural, forest, cropland, grassland pasture, and range

One primary way of studying drainage basin response is by creating hydrographs. A **hydrograph** is a graph showing variations in stream discharge over time (Figure 18.5). Examination of hydrographs for different rivers, or for the same river during two different time periods, provides a means of detecting differences in drainage basin response due to differences in characteristics such as soil type or land use.

The shape of the hydrograph indicates how the basin responds to storm events; it provides information on the speed and quantity of water moving through the basin to the mouth. The important characteristics of hydrograph shape are peak discharge, the slope of the rising and falling limbs, and the lag time. **Peak discharge** (Q_p) is the highest point on the hydrograph and this provides an indication of the severity of flood events. Drainage basins with low infiltration rates tend to have higher peak discharges than basins with high infiltration rates, and as a result, their potential flood hazard may also be greater. The **lag time** represents the length of time from the onset of a storm event to the peak discharge. Drainage basins with low infiltration rates and rapid runoff generally have short lag times. This is important in populated areas because the lag time indicates the amount of time available for evacuation prior to potential flooding. The **rising** and **falling limbs** of the hydrograph indicate the speed at which water levels rise and fall. Steep rising limbs are generally associated with short lag times and with drainage basins with characteristics that promote overland flow as opposed to infiltration. The falling limb indicates the speed at which excess storm water exits the basin. Storm water that has infiltrated the soil will move out of the basin slowly as throughflow, creating a gentle falling limb. In contrast, drainage basins with high rates of overland flow in which water quickly exits the basin will have a steep falling limb. Drainage basins with steep slopes, thin soils, or sparse vegetation tend to exhibit steep rising limbs, high peak discharges, and steep falling limbs. Drainage basins with gentle slopes, thick soils, or dense vegetation tend to exhibit gentle rising limbs, longer lag times, and gentle falling limbs.

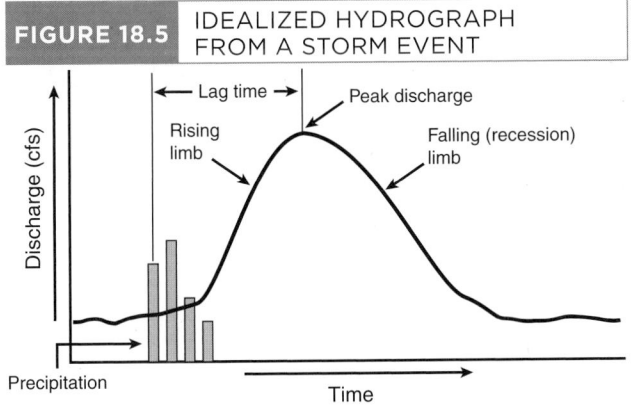

FIGURE 18.5 IDEALIZED HYDROGRAPH FROM A STORM EVENT

Floods and Flood Frequency

All rivers, at some time or another, will flood; flooding is a natural event for all rivers. When discharge increases to the point where it is at the brink of overflowing the banks of the stream channel, the stream is at *bankfull discharge*. If discharge increases above bankfull, a **flood** occurs—water spills over the river banks and floods the surrounding land. The land area outside of the stream channel covered with water is called the **floodplain.** The amount of land flooded depends on the **magnitude,** or size, of the flood event. Large floods will inundate a larger area of land than small floods. As a result, the definition of a floodplain for a small flood is not the same as the definition of a floodplain for a large flood.

Small floods occur fairly often. They do not do much damage and generally are not especially noticed. Large floods occur infrequently, do much damage, and receive much attention. Thus, the magnitude of a flood is inversely correlated with the **frequency** of a flood of that size. Examination of flood records for numerous rivers shows this to be true.

Since many communities are built along rivers, knowing the likelihood of various magnitude flood events is important. Given that there is a pattern to climate conditions year after year at a particular place, there also tends to be a pattern to flood events, although changes in drainage basin characteristics may affect the flood pattern independent of climate. Despite this, the magnitude and frequency of flood events for a particular river may be calculated, providing sufficient records are available. The data most often used for flood frequency analysis consists of the single highest flood event for each year for which records have been kept.

The **recurrence interval** or **return period** for a flood of a given magnitude is the average time within which a flood of that magnitude will be equaled or exceeded once. To calculate the return period, the flood events are ranked from largest to smallest. The recurrence interval for a particular discharge is calculated as:

$$\text{RI} = \frac{(n + 1)}{m} \quad (18.1)$$

where: RI = recurrence interval,

n = the number of years of record, and

m = the rank of the particular discharge.

Remember, the recurrence interval tells us, on the average, how often a flood of a particular magnitude occurs. The *probability* that a flood of a particular magnitude will be equaled or exceeded in a given year is:

$$P(\text{flood} \geq Q) = \frac{m}{(n + 1)} \quad (18.2)$$

where: P( ) = the probability of the event in parentheses,

Q = some selected discharge of interest,

n = the number of years of record, and

m = the rank of the particular discharge.

This probability can be converted to a percent, the percent of the time in which a given discharge is equaled or exceeded, by multiplying P(flood≥Q) by 100:

$$\% \text{ Time } Q_p \geq Q = \frac{100 \text{ m}}{(n+1)} \quad (18.3)$$

where: % Time $Q_p \geq Q$ = the percent of time in which peak discharge (Q_p) equals or exceeds some selected discharge of interest (Q),
n = the number of years of record, and
m = the rank of the particular discharge.

Many government agencies use these flood statistics to identify potential flood-hazard areas in communities located along rivers. The **hundred-year flood** is the flood magnitude that occurs on the average once every hundred years. The hundred-year floodplain is defined by the discharge magnitude for the hundred-year flood. The hundred-year flood is equaled or exceeded only one percent (1% or 0.01) of the time. Because many communities don't have flood records that go back one hundred years, flood-frequency analysis provides a tool for estimating the magnitude of that flood, based on the available records.

One very important point to remember when using and interpreting these flood statistics is that they represent averages. The statistics do not allow us to predict when a flood will occur; they tell us on the average how often floods of a given magnitude occur.

IMPORTANT TERMS, PHRASES, AND CONCEPTS

drainage basin
basin outlet or mouth
drainage divide
concave
convex
hydrologic cycle
overland flow/surface runoff
erosion
deposition
permeability
infiltration rate
interception
stream discharge

hydrograph
peak discharge
lag time
rising limb
falling limb
flood
floodplain
flood magnitude
flood frequency
recurrence interval/ return period
hundred-year flood

Name: _____   Section: _____

PART 1 · TOPOGRAPHY AND DRAINAGE BASINS

Use the topographic map of Thorofare Buttes, Wyoming (Appendix E, Figure E.16) to answer the questions. The stream network contributing water to Pass Creek and the stream network contributing water to Silvertip Creek have been highlighted on the map. This should allow you to see clearly that there are two separate stream networks on this map.

1. Mark in red, and label, the drainage divide that separates the streams draining into Pass Creek from the streams draining into Silvertip Creek. The drainage divide runs between these two sets of streams; it does not cut across any stream. The divide is defined by the highest points separating these two stream networks, so look for closed contour lines.

2. Four profile lines, A, B, C, and D, are marked on the Thorofare Buttes map (Appendix E, Figure E.16). Figure 18.6 contains the topographic profiles for these four lines. These topographic profiles show the slope shape in the *vertical* dimension. Use Figures E.16 and 18.6 to answer the following questions.

   a. Indicate whether the contour lines along each profile are closer together at the top of the slope or at the bottom of the slope—or whether they're evenly spaced along the slope. Then indicate the shape of the profile line as linear, concave, or convex.

	Spacing	**Profile Line Shape (Fig. 18.6)**
A – A	_____	_____
B – B	_____	_____
C – C	_____	_____
D – D	_____	_____

   b. Examine the spacing of the contour lines along lines E – E and F – F. What shape would a profile drawn along each of these lines have?

   E – E _____
   F – F _____

   c. What shape would a profile of the intermittent stream labeled "Z" have?

3. Determine the shape of each hillslope in the horizontal dimension (from one side to the other). First determine which end point has the highest elevation. Second, examine the *shape* of the contour lines as they cross the profile line and decide whether the contour lines bend uphill, bend downhill, or go straight across the profile line. Third, indicate the hillslope shape in the horizontal dimension as concave (a valley), convex (a ridge), or linear. Profile A has been done as an example.

	Highest End Point	**Direction Contour Lines Curve Toward**	**Hillslope Shape**
A – A	A	uphill	concave (valley)
B – B	_____	_____	_____
C – C	_____	_____	_____
D – D	_____	_____	_____
E – E	_____	_____	_____
F – F	_____	_____	_____
Stream Z	_____	_____	_____

FIGURE 18.6 TOPOGRAPHIC PROFILES, THOROFARE BUTTES, WYOMING. VERTICAL EXAGGERATION IS 5X

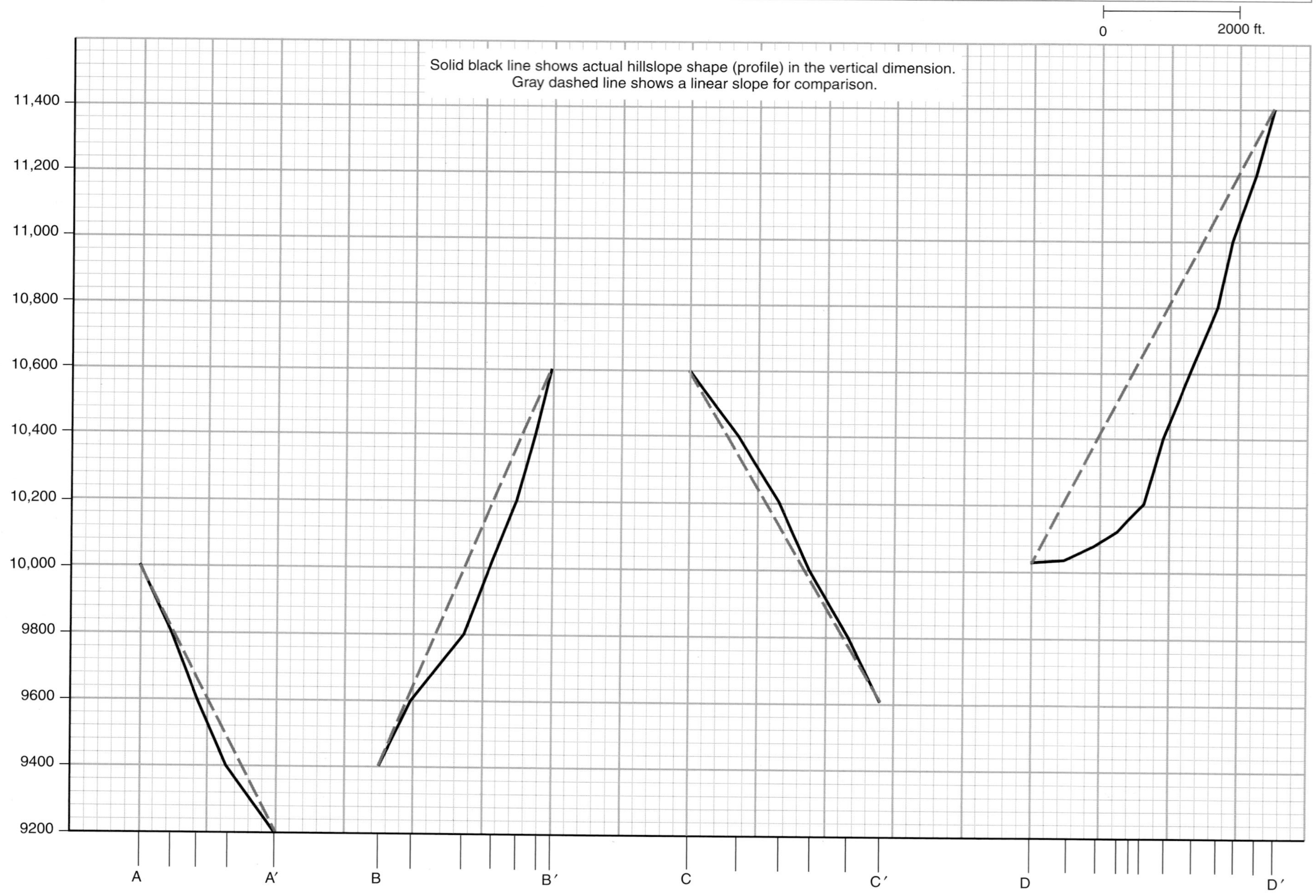
0
2000 ft.
Solid black line shows actual hillslope shape (profile) in the vertical dimension.
Gray dashed line shows a linear slope for comparison.
11,400
11,200
11,000
10,800
10,600
10,400
10,200
10,000
9800
9600
9400
9200
A
A′
B
B′
C
C′
D
D′

4. Based on your answers to question 3, which of the profile lines could be a drainage divide?

5. Combine your answers to questions 2 and 3 to determine which hillslope shape from Figure 18.3 matches each of the profile lines. Then indicate whether the slope will "gather water," or "spread water" or whether water will flow straight downhill. Last, indicate whether the erosive energy of water will be concentrated (occurs when water is gathered), spread out (occurs when water is spread), or evenly distributed (occurs when water flows straight downhill). Slope A – A has been done as an example.

	Hillslope Shape from Fig. 18.3	Water Flow	Erosive Energy
A – A	CC	gathered	concentrated
B – B			
C – C			
D – D			
E – E			
F – F			

6. a. If contour lines are bending uphill, do they define a valley or a ridge?

   b. If contour lines are bending downhill, do they define a valley or a ridge?

   c. If you want to locate a drainage divide, do you look for contour lines that are bending uphill or downhill?

   d. Given your answers to (a), (b), and (c), draw the drainage divide separating the tributary labeled "Y" from the tributary labeled "Z" on the Thorofare Buttes map (Appendix E, Figure E.16).

   e. Why is this drainage divide more difficult to locate at lower elevations than at higher elevations? What happens to the shape of the slope, as indicated by the shape of the contour lines, that makes it more difficult to locate?

   f. Is it really correct to draw this drainage divide all the way to Pass Creek as a single line? Why or why not?

PART 2 · SURFACE RUNOFF AND EROSION PROCESSES

1. Figure 18.7 shows an example of the relationship between interception storage, infiltration, and runoff during a storm event. Using Figure 18.7, what is the rate of:

	After 20 minutes	After 100 minutes
precipitation	_____	_____
interception	_____	_____
infiltration	_____	_____
surface runoff	_____	_____

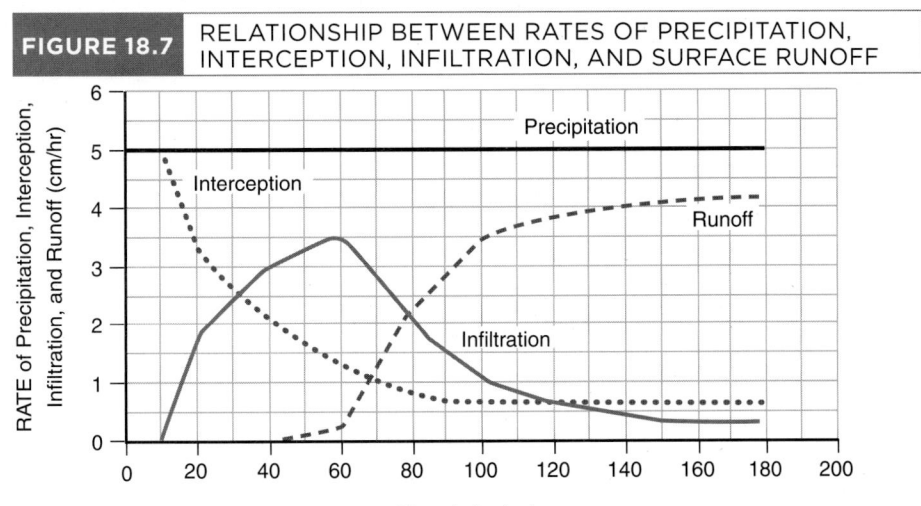

FIGURE 18.7 RELATIONSHIP BETWEEN RATES OF PRECIPITATION, INTERCEPTION, INFILTRATION, AND SURFACE RUNOFF

Note that the precipitation *rate* is a constant 5 cm/hour.

2. Assuming Figure 18.7 is representative of the relationship between precipitation, interception, infiltration, and surface runoff for most conditions, what generalizations can be made regarding the change over the course of a storm event in the rate of:

   a. interception

   b. infiltration

   c. surface runoff

231

3. Figure 18.7 is a rather simple figure for displaying a complex set of relationships. Identify two situations in which the relationships expressed in Figure 18.7 might not be completely true.

   a.

   b.

4. Since climate and vegetation patterns are related, there should be some correlation between rates of fluvial erosion, climate, and vegetation. Figure 18.8 shows the general relationship between runoff (a function of climate), erosion, and climate.

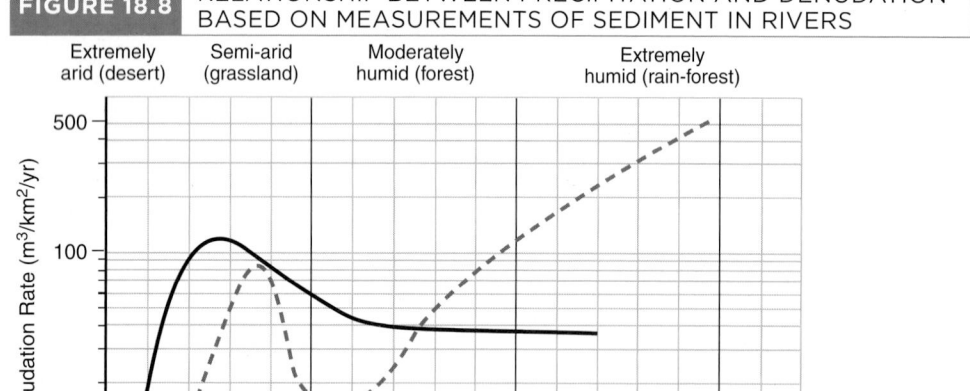

Source: Data after Langbein et al., 1958; and after Ohmori, 1983.

   a. What type of climate exhibits the highest erosion rate for the United States?

   b. What two types of climates produce the highest erosion rates for the world?

   c. What biomes (ecological communities) are associated with the climates you listed in parts (a) and (b) that produce high erosion rates?

   d. What two types of climates exhibit the lowest erosion rates for the United States and for the world?

   e. What biomes are associated with these climates producing low erosion rates?

f. Explain why these relationships exist. Keep in mind that climate and vegetation are closely related to one another, and that both affect rates of erosion.

5. Use Figure 18.9 to locate three areas of high and three areas of low erosion rates. Then determine the climate, natural vegetation, and topographic characteristics of these selected areas using Appendix F, Figures F.1, F.2, and F.3.

Regions of High Erosion	Climate	Natural Vegetation	Topography (gentle or steep)
_____	_____	_____	_____
_____	_____	_____	_____
_____	_____	_____	_____

Regions of Low Erosion	Climate	Natural Vegetation	Topography (gentle or steep)
_____	_____	_____	_____
_____	_____	_____	_____
_____	_____	_____	_____

FIGURE 18.9 WORLD DISTRIBUTION OF EROSION

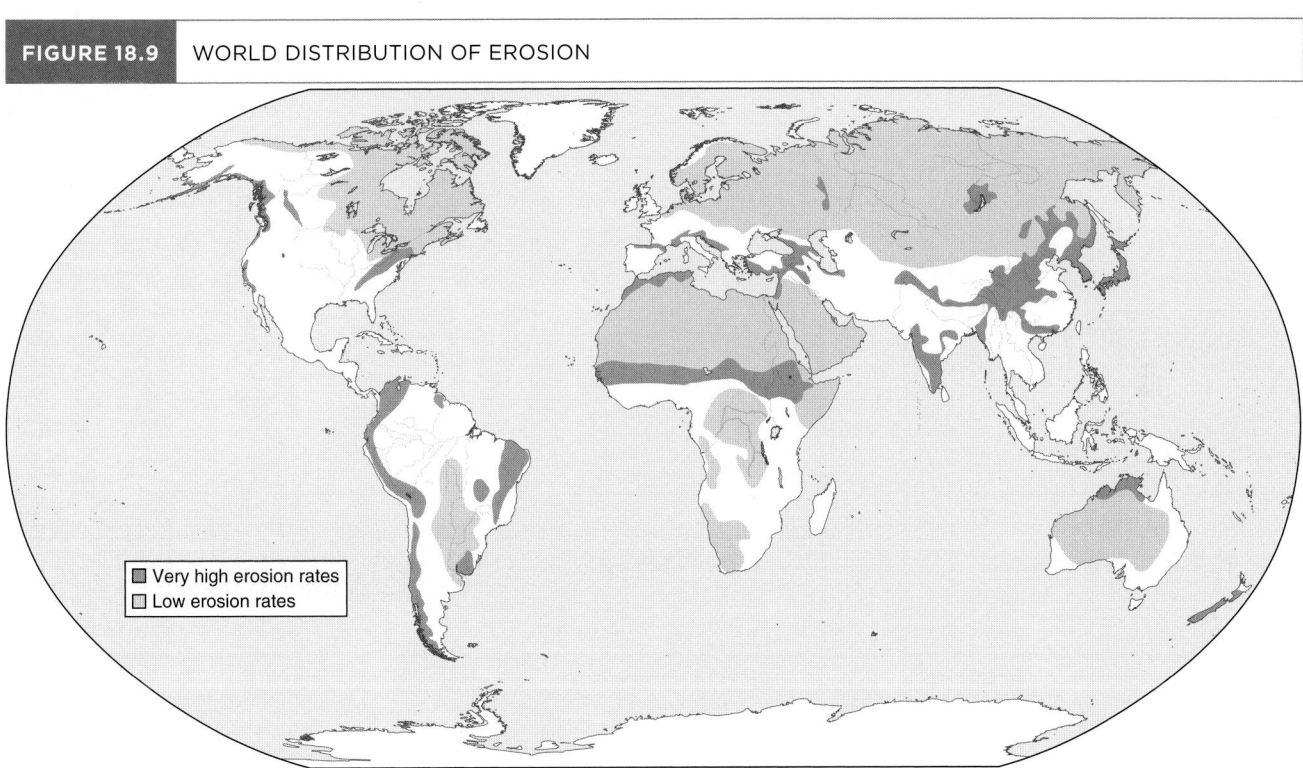

Areas not shaded have either intermediate amounts of erosion, or are areas of permanent ice cover or desert areas without measurements.

Source: Based on suspended sediment yield from Walling and Webb, 1983, Walling, 1987, Lvovich et al., 1991, and water erosion vulnerability from USDA, 2002.

6. a. Do the regions of highest and lowest erosion rates listed for question 5 have the types of climates and biomes you listed as answers to question 4? If not, which ones do not match?

   b. If the regions of high and low erosion you listed in question 5 do not have the types of climates and biomes you listed in question 4, explain what other factors might be contributing to high or low rates of erosion in these areas.

Name: _____   Section: _____

PART 3 • STORM HYDROGRAPHS

1. Table 18.3 contains data showing the response of two drainage basins to identical storm events (i.e., the storms produced identical amounts of rain at identical rates for identical lengths of time). These drainage basins are the same size but their characteristics (e.g., soils, vegetation, land use, topography) differ in some way. Plot the data in Table 18.3 on the graph paper to create a hydrograph for each drainage basin. The two hydrographs should be on the same graph, and they should begin at the same point in time. Label the axes clearly and label the hydrograph lines clearly.

TABLE 18.3 STREAMFLOW DATA FOR TWO DRAINAGE BASINS FROM IDENTICAL STORM EVENTS

| Drainage Basin A || Drainage Basin B ||
Time (hours)	Discharge (cfs)	Time (hours)	Discharge (cfs)
1.0	40	1.0	25
2.0	35	2.0	30
3.0	40	3.0	25
4.0	45	4.0	65
5.0	70	5.0	300
6.0	125	6.0	200
7.0	175	7.0	140
8.0	190	8.0	100
9.0	170	9.0	75
10.0	130	10.0	50
11.0	95	11.0	30
12.0	60	12.0	30
13.0	40	13.0	25
14.0	35	14.0	30
15.0	40	15.0	25
16.0	35	16.0	25

2. What is the peak discharge for:

   drainage basin A? _____ drainage basin B? _____

3. a. Which drainage basin experiences a longer lag time? _____

   b. How much longer is the lag time in this basin than in the other basin? _____

4. What can you deduce regarding the relative amounts of infiltration, overland flow, and throughflow (high, moderate, low) in each of these basins? Justify your answer in terms of the hydrograph shape (rising limb, peak discharge, falling limb).

   Drainage Basin A:

   infiltration

   overland flow

   throughflow

   Drainage Basin B:

   infiltration

   overland flow

   throughflow

5. Based on your deductions for question 4:

   a. how might soil characteristics differ in these two basins?

   b. how might vegetation characteristics differ in these two basins?

   c. how might topography differ in these two basins?

   d. What kinds of land use might produce a hydrograph similar to drainage basin A?

   e. What kinds of land use might produce a hydrograph similar to drainage basin B?

6. Which basin has a greater flood hazard risk? How do you know?

7. Based on the hydrographs, which of these two basins potentially experience higher erosion and sediment loads? How did you decide this?

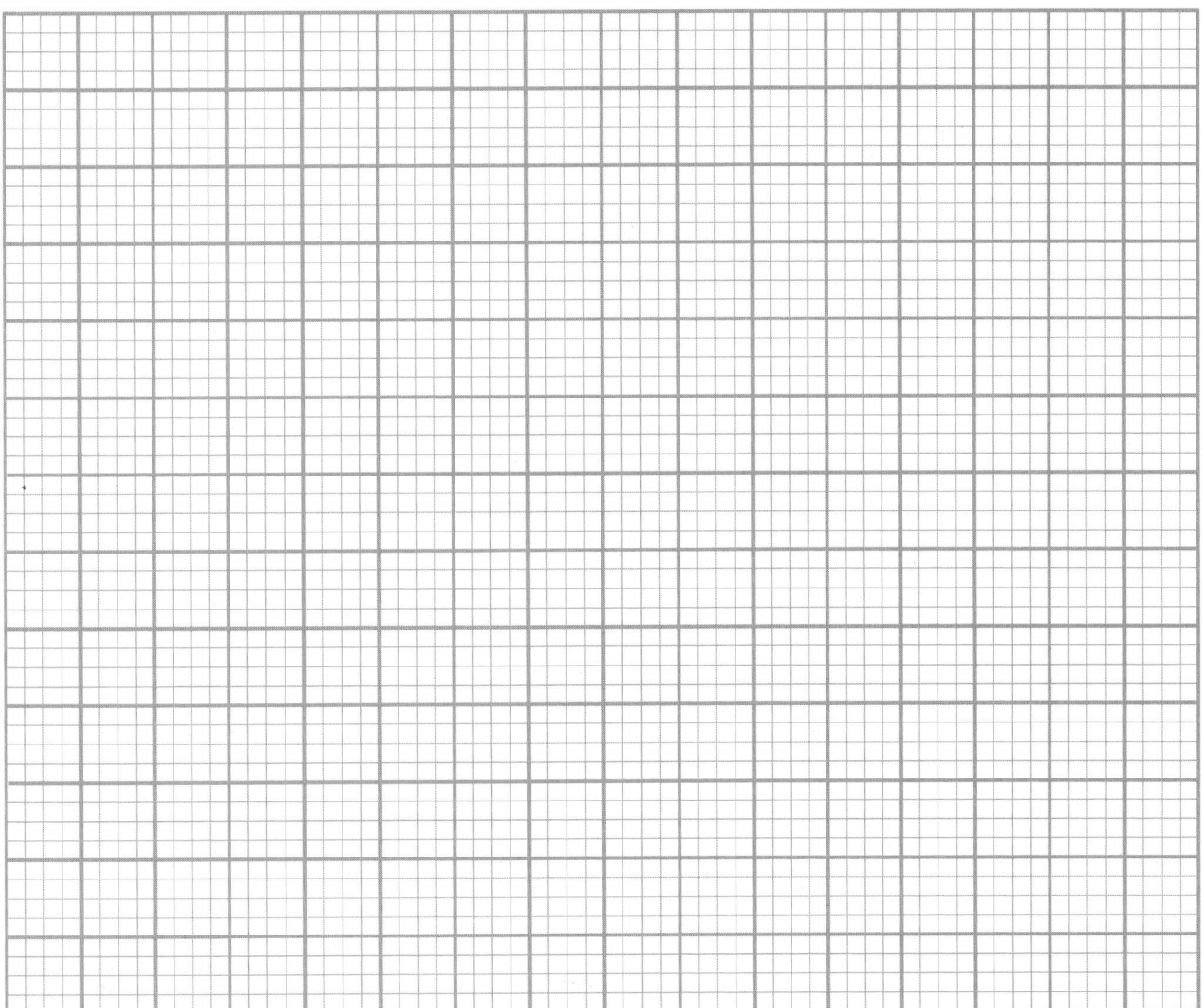

PART 4 • FLOOD MAGNITUDE AND FREQUENCY

Table 18.4 contains a record of the single highest peak discharge per year for an 84-year period (n = 84) on Jacks Fork at Eminence, Missouri. These discharges have been ranked in descending order of magnitude.

TABLE 18.4 — RANKED PEAK STREAMFLOW VALUES FOR JACKS FORK AT EMINENCE, MISSOURI, 1922-2005. SOURCE: USGS.

Rank	Year	Discharge (cfs)	% Time Q ≥ Q	Recurrence Interval (yrs)	Rank	Year	Discharge (cfs)	% Time Q ≥ Q	Recurrence Interval (yrs)
1	1994	58,500			20	1990	20,900	23.53	4.25
2	1986	55,800			21	1955	20,500	24.71	4.05
3	1985	51,500			22	1977	20,200	25.88	3.86
4	1983	50,300			23	2004	19,500	27.06	3.70
5	1966	46,200			24	1961	19,200	28.24	3.54
6	1993	44,100			25	1979	18,400	29.41	3.40
7	1974	40,600			26	1968	18,200	30.59	3.27
8	1928	40,000			27	1973	18,100	31.76	3.15
9	2002	35,900			28	1984	17,100	32.94	3.04
10	1969	32,500			29	1933	17,000	34.12	2.93
11	1943	31,000	12.94	7.73	30	1946	16,700	35.29	2.83
12	1950	31,000	14.12	7.08	31	1978	16,500	36.47	2.74
13	1935	29,500	15.29	6.54	32	1945	16,400	37.65	2.66
14	1999	27,100	16.47	6.07	33	1972	15,800	38.82	2.58
15	1949	27,000	17.65	5.67	34	1970	15,200	40.00	2.50
16	1956	27,000	18.82	5.31	35	1939	15,100	41.18	2.43
17	1963	24,200	20.00	5.00	36	1938	14,800	42.35	2.36
18	1975	22,900	21.18	4.72	37	1982	14,600	43.53	2.30
19	1957	21,600	22.35	4.47	38	1988	14,300	44.71	2.24

(continued)

TABLE 18.4 — RANKED PEAK STREAMFLOW VALUES FOR JACKS FORK AT EMINENCE, MISSOURI, 1922–2005. SOURCE: USGS. (CONTINUED)

Rank	Year	Discharge (cfs)	% Time Q ≥ Q	Recurrence Interval (yrs)	Rank	Year	Discharge (cfs)	% Time Q ≥ Q	Recurrence Interval (yrs)
39	1958	13,000	45.88	2.18	62	1930	7420	72.94	1.37
40	1960	13,000	47.06	2.13	63	1922	7240	74.12	1.35
41	1997	12,500	48.24	2.07	64	1965	7100	75.29	1.33
42	1923	12,200	49.41	2.02	65	1991	7010	76.47	1.31
43	1964	12,200	50.59	1.98	66	1971	5680	77.65	1.29
44	1989	12,100	51.76	1.93	67	2003	5260	78.82	1.27
45	1992	11,600	52.94	1.89	68	1925	5070	80.00	1.25
46	1981	11,000	54.12	1.85	69	1940	4450	81.18	1.23
47	1927	10,900	55.29	1.81	70	1976	4360	82.35	1.21
48	1959	10,300	56.47	1.77	71	1926	4270	83.53	1.20
49	1951	9860	57.65	1.73	72	1953	4150	84.71	1.18
50	1947	9640	58.82	1.70	73	1954	3400	85.88	1.16
51	1998	9450	60.00	1.67	74	1987	3290	87.06	1.15
52	1995	9350	61.18	1.63	75	1924	2970	88.24	1.13
53	1948	8960	62.35	1.60	76	2000	2860	89.41	1.12
54	1952	8870	63.53	1.57	77	1931	2740	90.59	1.10
55	1962	8620	64.71	1.55	78	1980	2650	91.76	1.09
56	1929	8360	65.88	1.52	79	1936	2620	92.94	1.08
57	1942	8050	67.06	1.49	80	1932	2610	94.12	1.06
58	1937	7820	68.24	1.47	81	1944	2570	95.29	1.05
59	2001	7810	69.41	1.44	82	1967	2320	96.47	1.04
60	2005	7620	70.59	1.42	83	1941	1860	97.65	1.02
61	1996	7480	71.76	1.39	84	1934	1270	98.82	1.01

1. a. Using equation 18.1, calculate the missing values for recurrence intervals (return periods) and add this information to Table 18.4. Round all answers to two decimal places.

   b. Using equation 18.3, calculate the missing values for the percent of time that peak discharges are equaled or exceeded (% Time $Q_p \geq Q$) and add this information to Table 18.4.

2. a. Figure 18.10 is a plot of discharge against the recurrence interval. Add points to this graph for the values you calculated for question 1.

   b. Once the points are plotted, draw a *trend line* showing the general pattern of discharge frequency. Remember, a trend line is a line that is centered on the dots, not a line that connects the dots; some of the dots will fall above the line and some below it. In this particular case, your trend line will not be perfectly straight; it will have a bit of a curve to it. The line should cover the entire width of the graph (from 1.01 to 200 years).

3. a. What is the largest annual peak discharge (Q_p) recorded? _____

   b. On the average, how often does annual peak streamflow get this high?

   c. What is the smallest annual peak discharge (Q_p) recorded? _____

   d. On the average, how often does annual peak streamflow get this high?

   e. Based on your answers to parts (a) through (d), and based on the graph in Figure 18.10, what relationship exists between flood magnitude and flood frequency?

4. Every time the flow on this river exceeds 46,000 cfs a small town on the river floodplain gets flooded. Approximately how often does this town get flooded (years)?

5. In the year 2000, the highest flow was 2860 cfs, and according to Table 18.4, this flood has a return period of approximately one year.

   a. The return period implies that a discharge of 2860 cfs occurs on the average once every year; however, is it possible that a flood of this magnitude could have occurred two or three times in 2000? Why or why not?

   b. What percent of the time is a peak streamflow of 2860 cfs equaled or exceeded?

   c. The largest flood on record was 58,500 cfs and occurred in 1994. This flood has an 85-year return period. Knowing this, can we predict when a flood of this magnitude will occur again? Why or why not?

   d. What percent of the time is a peak streamflow of 58,500 cfs equaled or exceeded?

FIGURE 18.10 FLOOD FREQUENCY FOR JACKS FORK AT EMINENCE, MISSOURI

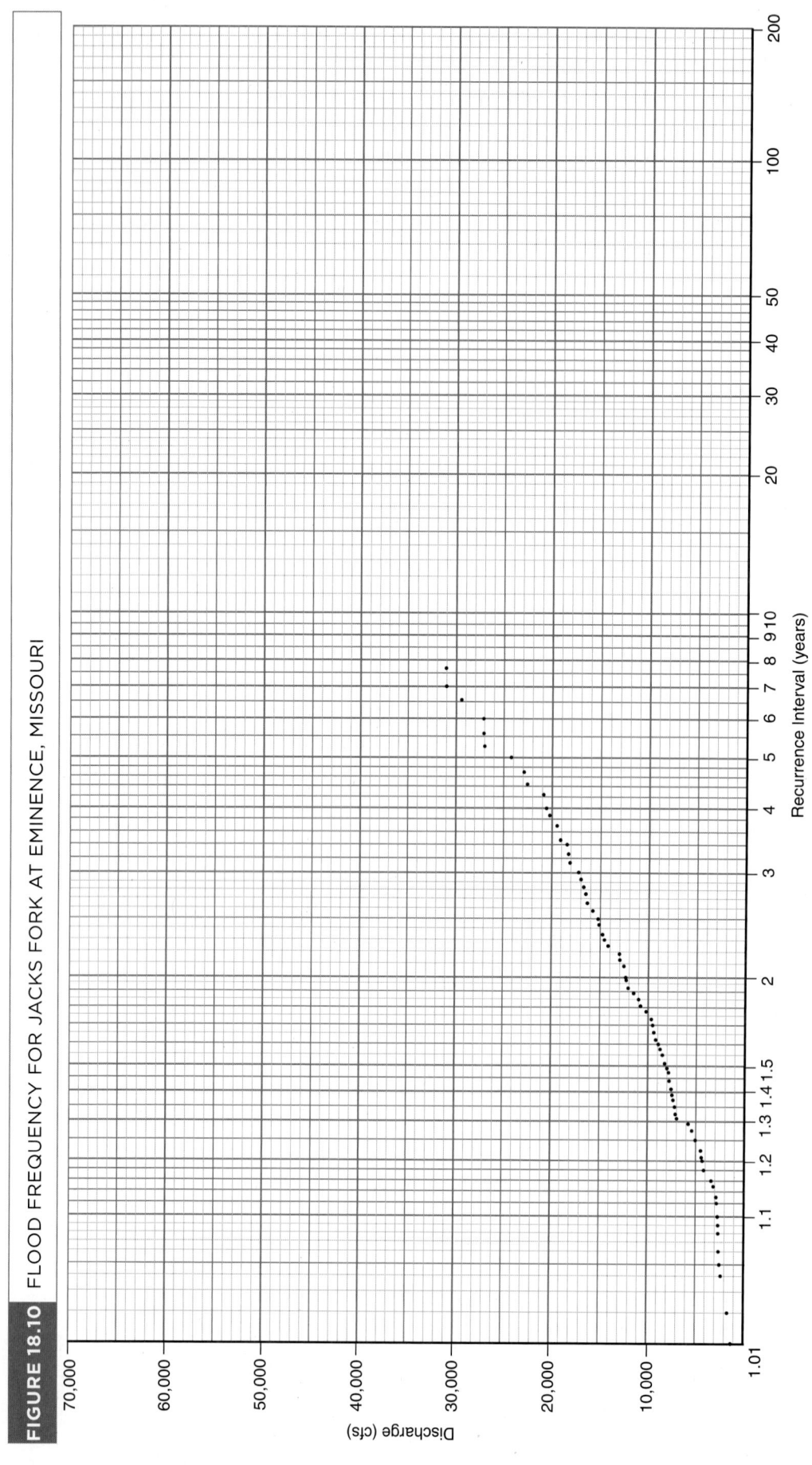

Source: Data from USGS.

6. a. The 50-year flood is defined as the flood that occurs, on the average, once every 50 years (i.e., the flood with a 50-year return period). What percent of the time will the 50-year flood be equaled or exceeded?

b. Based on your trend line in Figure 18.10, what is the discharge for the 50-year flood?

c. The 100-year flood is the flood magnitude with a return period of 100 years. What percent of the time will the 100-year flood be equaled or exceeded?

d. The plotted points on the graph in Figure 18.10 do not include a point for the 100-year flood. Why not?

e. Based on your trend line in Figure 18.10, what is the estimated discharge for the 100-year flood?

f. Assume the 100-year flood has a stage (the elevation of the water surface) of 634 feet above sea level. If you found the perfect spot for your dream house, perfect in every way except for the fact that it has an elevation of 630 feet above sea level, would the risk of flooding be great enough to deter you from building? Why or why not?

7. Suppose an annual peak flood (Q_p) with a magnitude of 62,000 cfs occurred in Eminence.

a. Based on your trend line in Figure 18.10, what is the return period for this flood?

b. Knowing the return period, what percent of the time can discharge be expected to equal or exceed 62,000 cfs? (%time $Q_p \geq Q = 1/R_I$)

c. Can we predict when this flood will occur again? Why or why not?

EXERCISE 20 ⭐ GLACIAL LANDFORMS

PURPOSE

The purpose of this exercise is to learn how to identify glacial landforms and to examine the processes that create these landforms.

LEARNING OBJECTIVES

By the end of this exercise you should be able to

- explain the relationship between mass balance of alpine glaciers, equilibrium line elevation, and altitude;
- determine whether a glacier is advancing or retreating, based on net mass balance, and make inferences about past glacial extent, based on landform information;
- identify alpine and continental glacial landforms on topographic maps and air photos; and
- describe some of the characteristics of glacial landforms and describe where we would find these landforms.

INTRODUCTION

Glaciers form in areas where the climate is cold enough for winter snowfall to not melt in the summer season. **Alpine glaciers** form in mountainous regions under current climatic conditions. As altitude increases, temperature decreases, allowing snow to accumulate from one year to the next. **Continental glaciers** exist today at high latitudes, such as in Antarctica and Greenland, where temperatures are perennially cold. These glaciers are much larger in size than alpine glaciers. Continental glaciers also covered large portions of North America and Northern Europe and Asia during the last ice age when global climatic conditions were cooler than they currently are. As glacial ice flows across the earth's surface, it modifies the landscape through the processes of erosion, transportation of material, and deposition of material, creating a variety of erosional and depositional landforms.

Glacial Mass Balance

In order for glaciers to form and survive, the prevailing climate must allow the accumulation of snow to exceed ablation. Glacial mass balance involves determining the overall gains or losses of a glacier's mass and helps determine whether a glacier is growing or shrinking. **Accumulation** of snow, which eventually transforms to ice, occurs on the upper portion of alpine glaciers in the accumulation zone (Figure 20.1). **Firn** is snow that has survived through the summer melt season and is intermediate between snow and ice. Firn is also found in the accumulation zone. **Ablation** refers to all the losses sustained through evaporation, melting, or chunks of ice breaking off the glacier. Ablation dominates in the lower portion of alpine glaciers; that portion is called the ablation zone (Figure 20.1). In the ablation zone we find only ice, thus the **firn line** marks the lowest elevation where firn is found. The firn line also indicates the border between the accumulation and ablation zones.

The **mass balance** of a glacier is equal to the gain of snow and ice minus the loss of snow and ice over the balance year. The **balance year** does not necessarily correlate with a calendar year; it is based on seasonal changes in accumulation and ablation patterns. The **winter season** begins when accumulation exceeds ablation, generally in late summer or early fall, and as a result, the winter balance is positive. The **summer season** begins when ablation exceeds accumulation, usually in late spring, and the summer balance is negative. The net balance over the course of a year equals the summer balance plus the winter balance:

$$B_n = B_s + B_w \qquad (20.1)$$

where B_n is the net balance, B_s is the summer balance, and B_w is the winter balance. Figure 20.2 shows graphically the difference in the summer, winter, and net balances. Note that B_w is positive at all altitudes and B_s is negative at all altitudes.

FIGURE 20.1 GLACIAL ACCUMULATION AND ABLATION ZONES

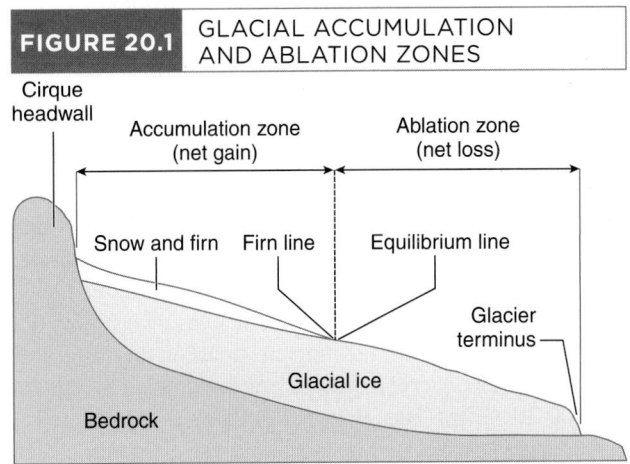

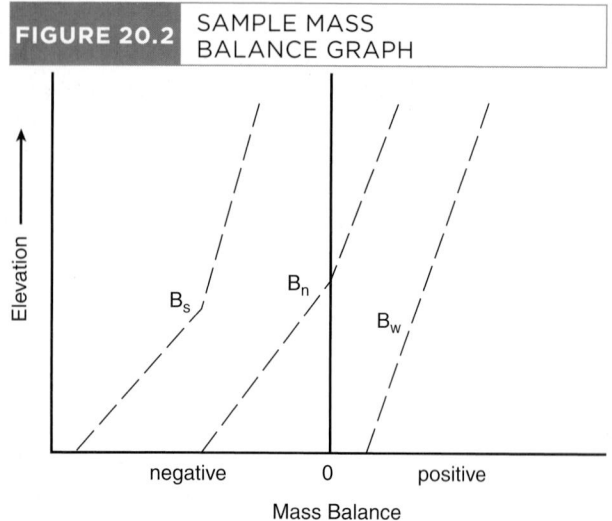

FIGURE 20.2 SAMPLE MASS BALANCE GRAPH

Net mass balance (B_n) is positive in the accumulation zone at higher elevations and negative in the ablation zone at lower elevations.

Numerous measurement sites are necessary to determine a glacier's mass balance. These sites are often situated at different elevations along a glacier. The mass balance is calculated for each measurement site in water equivalency depths. The **equilibrium line** is an imaginary line on the surface of a glacier that marks the elevation where accumulation and ablation are equal: the net mass balance is zero. The equilibrium line divides the accumulation zone from the ablation zone (Figure 20.1). The elevation of the equilibrium line is influenced by prevailing local climatic conditions, in particular latitude and glacier orientation, as well as net mass balance. Advancing glaciers tend to have lower equilibrium line elevations than retreating glaciers at the same latitude.

If the net mass balance for the entire glacier is positive, the glacier will grow, causing the terminus to extend away from the source region. That is, the glacier will **advance.** If the net mass balance for the entire glacier is negative, the glacier will shrink, causing the terminus to melt back toward the source. That is to say, the glacier will **retreat.** If the net mass balance for the entire glacier is zero, the terminus remains **stationary** on the landscape: the glacier doesn't grow or shrink. Regardless of whether an alpine glacier is advancing, retreating, or stationary, the ice is always flowing downhill; in alpine environments ice does not flow uphill! Retreating alpine glaciers are simply melting faster than they're flowing forward.

As glacial ice flows across the landscape, it doesn't always flow in a uniform fashion. In alpine areas, the slope of the landscape underlying the glacier affects the flow rate. In steeper areas, the ice flows faster than in flatter areas. Because ice cannot deform rapidly, in areas of fast flow the ice tends to break apart, creating crevasses on the surface of the glacier. These **crevasses** are cracks that can extend as much as 30 meters (100 ft) down from the surface. As the ice flows, new crevasses will open and old crevasses will close.

Alpine Glacial Landforms

As glaciers form, advance across the landscape, retreat, and disappear, a distinct array of landforms is created. The landforms of mountainous regions that have not undergone glaciation are different from the landforms of mountainous regions that have been glaciated. Once the glaciers are gone, the landforms created by the glaciers remain as a reminder of that location's past climatic condition. Over thousands and millions of years, these glacial landforms are eroded and altered, providing clues as to how long ago glaciers existed.

Figure 20.3 shows landforms found during glaciation and after glaciers have melted. In unglaciated mountainous

FIGURE 20.3 LANDFORMS FOUND DURING AND AFTER ALPINE GLACIATION

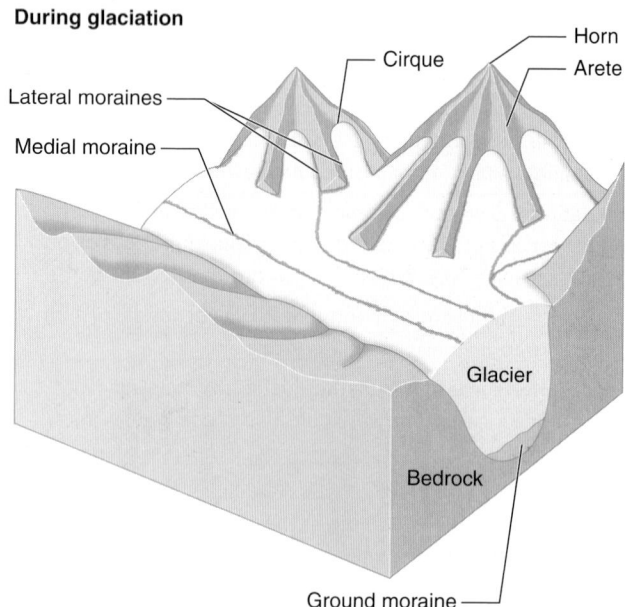

During glaciation

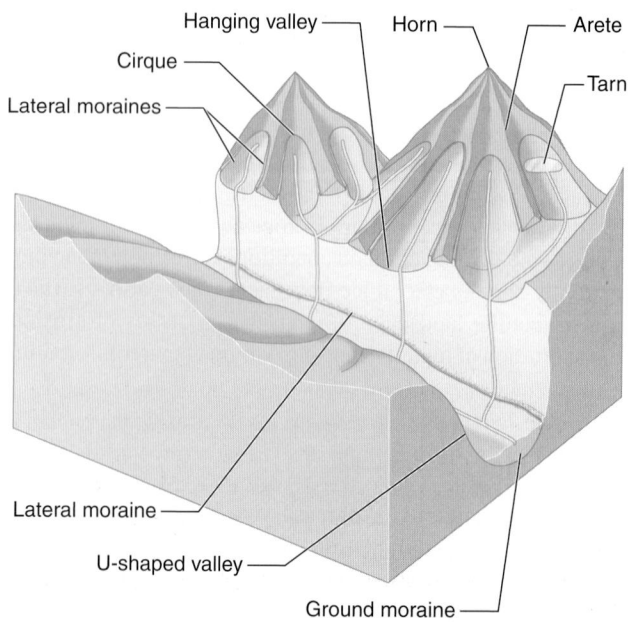

After glaciation

256

regions, river valleys have narrow **V-shaped** profiles with narrow valley bottoms. Since the valleys were created by rivers, the valleys may meander. Regions that have been glaciated are characterized by wide **U-shaped** or *parabolic* valley profiles with wide, relatively flat valley floors and steeper sides. Glaciers tend to straighten out valleys as they erode them, thus glacially eroded U-shaped valleys tend to be straighter than fluvially eroded V-shaped valleys. Small tributary streams join larger rivers at the same elevation in nonglaciated regions, but in glaciated regions tributary streams do not join larger rivers at the same elevation. This occurs because glaciers erode the floor of main stream valleys more than small tributary stream valleys, leaving the tributary valleys high above the elevation of the main valley floor as **hanging valleys,** ideal locations for waterfalls.

Glacial erosion of the landscape results in a number of other unique landforms. **Cirques** are semicircular, bowl- or amphitheater-shaped bedrock features formed where the glacial ice accumulates. Cirques are the source areas for glacial ice. As cirques fill with snow and ice, the glacial ice erodes back into the mountain, producing semicircular-shaped depressions. Once the ice has melted, glacial lakes called **tarns** often form in the bottom of cirques. **Horns** are pointed or pyramid-shaped mountain peaks formed when two or more glaciers erode back into the same mountain. **Arêtes** are sharp-edged bedrock ridges formed between two cirques or two valley glaciers as glaciers erode away the sides of the ridge.

As glaciers erode the landscape, they pick up debris and carry it away. The debris may range in size from very fine silt-sized particles, called *glacial flour,* to sand, gravel, pebbles, rocks, and large boulders. Eventually, this material is deposited on the landscape, creating depositional landforms. **Moraines** consist of accumulations of unconsolidated sediments and debris deposited by glacial ice. They may form on top of glaciers, in front of glaciers, along the sides of glaciers, or under glaciers.

Lateral moraines are ridges of material that accumulates along the sides of valley glaciers. Due to constant freeze-thaw weathering of the valley walls, debris continually falls on top of glaciers along the valley sides, creating these lateral moraines. When two or more valley glaciers join, their lateral moraines merge to form a **medial moraine** extending down the middle of the glacier. Both lateral and medial moraines can be seen as ridges of darker debris on top of existing glaciers. Once these glaciers melt, lateral moraines may remain behind, but medial moraines are not easily distinguished on the landscape. **End moraines** are ridges of debris that form in front of the terminus of a glacier. In alpine areas, end moraines often stretch across the valley the glacier filled. **Terminal moraines,** a type of end moraine, mark the farthest extent of glacial advance, thus these moraines are located at lower elevations. As glaciers retreat, **recessional moraines,** another type of end moraine, mark locations where the glacier paused before retreating farther upvalley. Recessional and terminal moraines are essentially alike, except that recessional moraines are located at higher elevations behind the terminal moraine. **Ground moraine** consists of debris deposited directly beneath the glacial ice.

Continental Glaciers

During the most recent ice age, the climate for much of the earth was cooler than it is today. This allowed continental glaciers to grow and eventually cover large areas of the earth's surface in the northern hemisphere. A number of glacial advances and retreats of these continental glaciers occurred throughout this ice age in response to fluctuations in the earth's temperature. The last advance of the continental ice sheets in North America is known as the Wisconsin advance. The glaciers from this period retreated approximately 10,000 years ago.

Three general terrain types associated with continental glaciers include areas of end moraines, areas of ground moraine, and outwash plains (Figure 20.4). End moraines are uneven ridges or mounds of debris deposited at the terminus of the glacier, and as with alpine glaciers, include terminal and recessional moraines. Ground moraine from continental glaciers, as with alpine glaciers, consists of debris deposited

FIGURE 20.4 LANDFORMS FOUND DURING AND AFTER CONTINENTAL GLACIATION

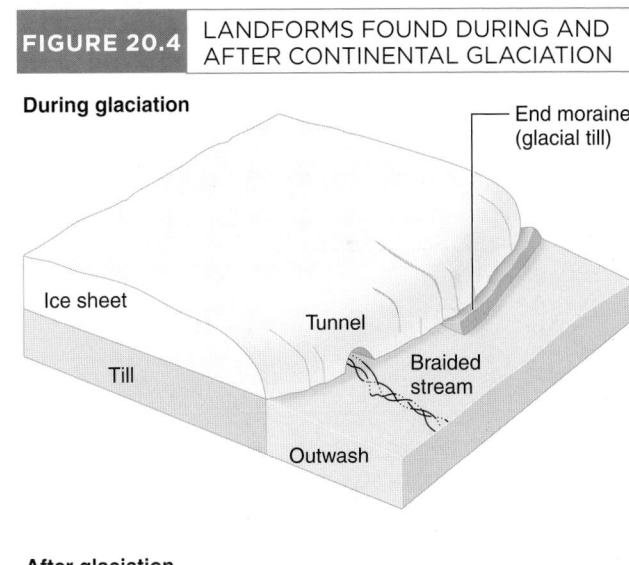

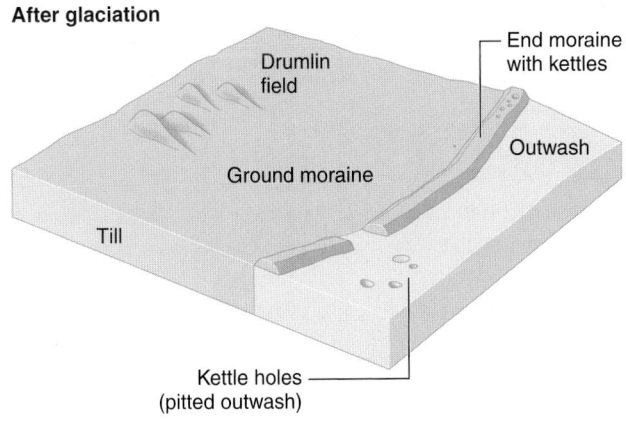

directly beneath the glacial ice. Ground moraine is located behind end moraines. In areas of continental glaciers, once ground moraine is uncovered, it may be quite poorly drained with marshy areas present. **Drumlins** are elongate mounds or hills that form beneath the glacial ice and are found in areas of ground moraine. These mounds are made up of sediments molded and compressed into long ridges. The long axis of these hills or ridges is oriented parallel to the direction of ice flow. The end of the mound from which the glacier advanced is generally steeper than the "down-glacier" end. This can be seen by the spacing of contour lines on topographic maps. One other type of moraine associated with continental glaciers is an **interlobate moraine.** As ice in continental ice sheets spreads across the landscape, lobes form. Ice spreads outward from the lobes, carrying debris with it. When two lobes abut one another, debris tends to accumulate along the border zone, forming an interlobate moraine.

Outwash plains are relatively flat areas stretching out beyond an end moraine. These plains are created as meltwater from the glacier washes sand and other sediments out of the glacial ice and deposits these across a broad flat region. A **pitted outwash plain** also appears as a broad plain stretching out ahead of an end moraine, but the plain is dotted with depressions creating "pitted topography." The pitted topography of the outwash plain is due to the presence of **kettle holes.** Kettle holes form as blocks of ice disconnected from the main body of the glacier melt and allow the overlying sediment to collapse, forming a depression.

IMPORTANT TERMS, PHRASES, AND CONCEPTS

- alpine glaciers
- continental glaciers
- accumulation
- firn
- ablation
- firn line
- mass balance
- balance year
- winter and summer seasons
- equilibrium line
- advancing glaciers
- retreating glaciers
- stationary glaciers
- crevasses
- V-shaped valleys
- U-shaped valleys
- hanging valleys
- cirques
- tarns
- horns
- arêtes
- moraines
- lateral moraines
- medial moraines
- end moraines
- terminal moraines
- recessional moraines
- ground moraine
- drumlins
- interlobate moraines
- outwash plains
- pitted outwash plains
- kettle holes

Name: _____   Section: _____

PART 1 ✦ ALPINE GLACIAL PROCESSES AND LANDFORMS

1. a. On a topographic map, how can you distinguish the contour lines representing glaciers from those representing solid bedrock or bare ground?

   b. On topographic maps, how can you distinguish glacial moraines, which are composed of till (unconsolidated sediments) from solid bedrock or bare ground?

2. Examine the Mount Rainier topographic map (Appendix E, Figure E.19) in conjunction with the stereo triplet of Mount Rainier (Figure 20.5). On the stereo triplet, you can see only two of these photos in three dimensions at one time; you can't see all three at once. Refer to Appendix D for information on using stereoscopes to view air photos in three dimensions.

   a. How can you distinguish forested areas from nonforested areas on topographic maps?

   b. How can you distinguish forested areas or areas of bare ground from ice- or snow-covered areas on the air photos?

   c. On the air photos, why are the glaciers more difficult to define at lower elevations than at higher elevations? Examination of the topographic map might help answer this question.

FIGURE 20.5 STEREO TRIPLET OF MOUNT RAINIER, 1955

Source: Photos used with permission of American Educational Products, Ft. Collins, CO.

3. Use the Mount Rainier topographic map (Appendix E, Figure E.19) and the two photos on the left of Figure 20.5 to answer the following questions. Note the direction of the north arrow on the two figures. Rotate your map so it matches the orientation of the air photos.

   a. Find an example of a horn on the air photos and circle it.

   What is the name of this horn on the topographic map? _____

   How many glaciers surround this horn? _____

   b. Are horns erosional or depositional landforms? Explain.

   c. There are several well-defined arêtes on the air photos. Locate and circle two of them.

   If these arêtes are named on your map, what are their names? If they are not named, what glaciers are they located between?

   d. What are two other examples of arêtes on the topographic map?

   e. Examine the four arêtes you just identified on the topographic map. How would you describe the general location of arêtes in glacial landscapes?

   f. If you had to draw a topographic profile going across an arête, what would your profile look like?

   g. Are arêtes erosional or depositional landforms? Explain how this relates to their shape.

   h. Are horns and arêtes composed of solid bedrock or glacial till? How do you know this from looking at the topographic map?

4. Examine the stereo air photos of Carbon Glacier (Figure 20.6), along with the Mount Rainier topographic map (Appendix E, Figure E.19).

   a. Locate and circle or highlight Willis Wall, the steep slope found surrounding the start of the accumulation zone of Carbon Glacier, on the topographic map and the air photos.

FIGURE 20.6 STEREO AIR PHOTOS OF CARBON GLACIER, 1955

Source: Photos used with permission of American Educational Products, Ft. Collins, CO.

b. Is Willis Wall composed of bedrock or is it composed of glacial till? How do you know?

c. If you had to describe the appearance of Willis Wall to a friend who had never seen it, how would you describe its shape?

d. What type of landform is Willis Wall? _____

e. Is Willis Wall an erosional or a depositional landform? Explain.

f. At its left and right sides, Willis Wall grades into another glacial landform. What type of landform is this?

5. a. Examine the slope of the surface of Carbon Glacier on the air photos (Figure 20.6). There are three locations where the slope of the glacier surface is quite steep relative to the rest of the glacier surface. Circle these three locations on the air photo.

   b. What features do you see on the glacier surface at these three locations?

   c. Why are these features evident here and not elsewhere?

6. Examine the moraine deposits covering the surface of Carbon Glacier near the terminus on the topographic map (Appendix E, Figure E.19) and the air photos (Figure 20.6).

   a. Do the moraine deposits covering the glacier surface extend as far uphill on the air photos as on the topographic map?

   b. Explain why this might be the case.

7. Examine the lower portion of Ingraham and Cowlitz Glaciers on the Mount Rainier topographic map (Appendix E, Figure E.19).

   a. What type of moraine is shown just downstream from where these two glaciers merge? Why is this moraine forming here?

   b. Slightly downhill from where these glaciers merge, moraines are shown along the valley walls. What type of moraines are these, and why are they forming here?

8. Moraine deposits can provide information on the past extent of glaciers and their mass balance. Examine the moraine deposits at the short line labeled "A" near the terminus of Cowlitz Glacier on the Mount Rainier topographic map (Appendix E, Figure E.19). Line A extends from the current edge of the glacier to the edge of the moraine deposits.

   a. What is the elevation of the edge of the glacier at line A? _____

   What is the elevation of the top edge of the moraine deposits? _____

   What might this difference in elevation indicate about the thickness of the ice at this location today compared to the thickness of the ice at this location in the past?

   b. What is the approximate elevation of the terminus of Cowlitz Glacier?

   What is the approximate lowest elevation where we find moraine deposits downstream from Cowlitz Glacier?

   What might this difference in the elevation indicate about the current length of Cowlitz Glacier compared to its length in the past?

c. Figure 20.7 shows a profile across the Cowlitz River, the river that drains the Cowlitz Glacier. This profile is located approximately 4 miles downstream from the edge of your map. The topographic map shows no brown speckles indicating moraine deposits at the location of this profile. Based on the appearance of this profile, do you think the Cowlitz Glacier may have extended this far downvalley some time in the past? Why or why not?

d. Based on your answers to parts (a), (b), and (c), how would you classify the overall net mass balance (B_n) of Cowlitz Glacier?

e. Examine Emmons Glacier. Keeping in mind your answers to the previous questions, what do you think the mass balance to Emmons Glacier is? What evidence did you use to determine this?

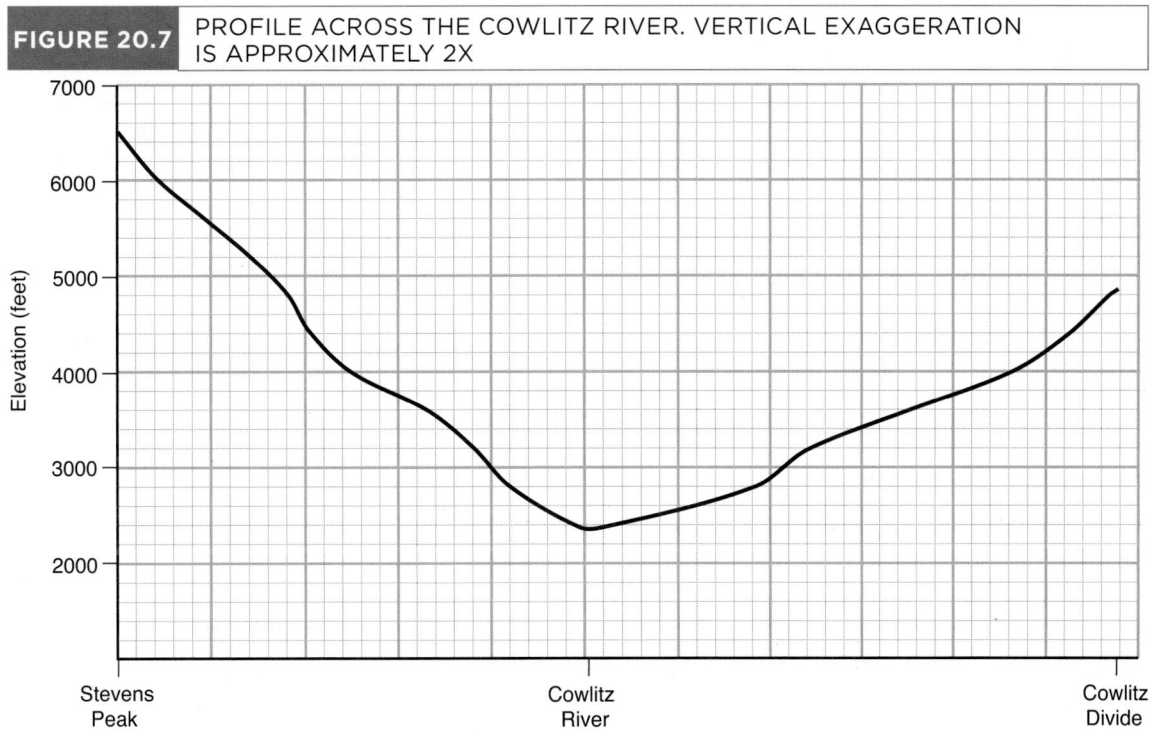

FIGURE 20.7 PROFILE ACROSS THE COWLITZ RIVER. VERTICAL EXAGGERATION IS APPROXIMATELY 2X

9. Data on snow accumulation and ablation for Nisqually and Emmons Glaciers for 2003 are presented in Table 20.1. Some of these values are direct measurements made by the National Park Service, while others are estimated values based on models of how ablation and accumulation change with altitude. The National Park Service must estimate the values for locations on the two glaciers that are not accessible for direct measurements.

   a. Calculate the net mass balance at each elevation for each glacier and enter the information into Table 20.1. Remember, $B_n = B_s + B_w$.

   b. On the graph in Figure 20.8, plot net mass balance against altitude for each glacier. Connect the points for each glacier and label the two lines clearly.

   c. What is the general relationship between altitude and net mass balance as shown in Figure 20.8?

   d. Based on Figure 20.8, what is the elevation of the equilibrium line for each glacier?

   Emmons Glacier _____

   Nisqually Glacier _____

   e. The average net mass balance represents the average thickness of an evenly thick layer of water lost (if the value is negative) or gained (if the value is positive) over the entire surface of the glacier (measured in units of water equivalent). Determine the average net mass balance for each glacier by summing the values for B_n and dividing by the number of observations.

   Emmons Glacier _____

   Nisqually Glacier _____

   f. Assuming these two glaciers are representative of glaciers in general, what is the relationship between the average net mass balance and the elevation of the equilibrium line?

   g. If a glacier continually shrinks over several decades, what will happen to the elevation of the equilibrium line? Why?

   h. Which of these two glaciers has a more negative net mass balance? _____

   What might be an explanation for why this is the case?

TABLE 20.1 — GLACIAL MASS BALANCE DATA (UNITS IN INCHES WATER EQUIVALENT)

Elevation (feet)	Emmons Glacier B_s	Emmons Glacier B_w	Emmons Glacier B_n	Nisqually Glacier B_s	Nisqually Glacier B_w	Nisqually Glacier B_n
14,500	0	+100		−25	+70	
13,500	0	+90		−55	+70	
12,500	−50	+85		−80	+70	
11,500	−80	+80		−120	+70	
10,500	−135	+80		−135	+70	
9500	−185	+90		−150	+70	
8500	−225	+100		−175	+100	
7500	−213	+110		−250	+125	
6500	−250	+75		−235	+115	
5500	−300	+30		−360	+60	

Source: Data from National Park Service (2004) http://www.nps.gov/mora/ncrd/update2004.doc

FIGURE 20.8 — RELATIONSHIP BETWEEN NET MASS BALANCE AND ALTITUDE

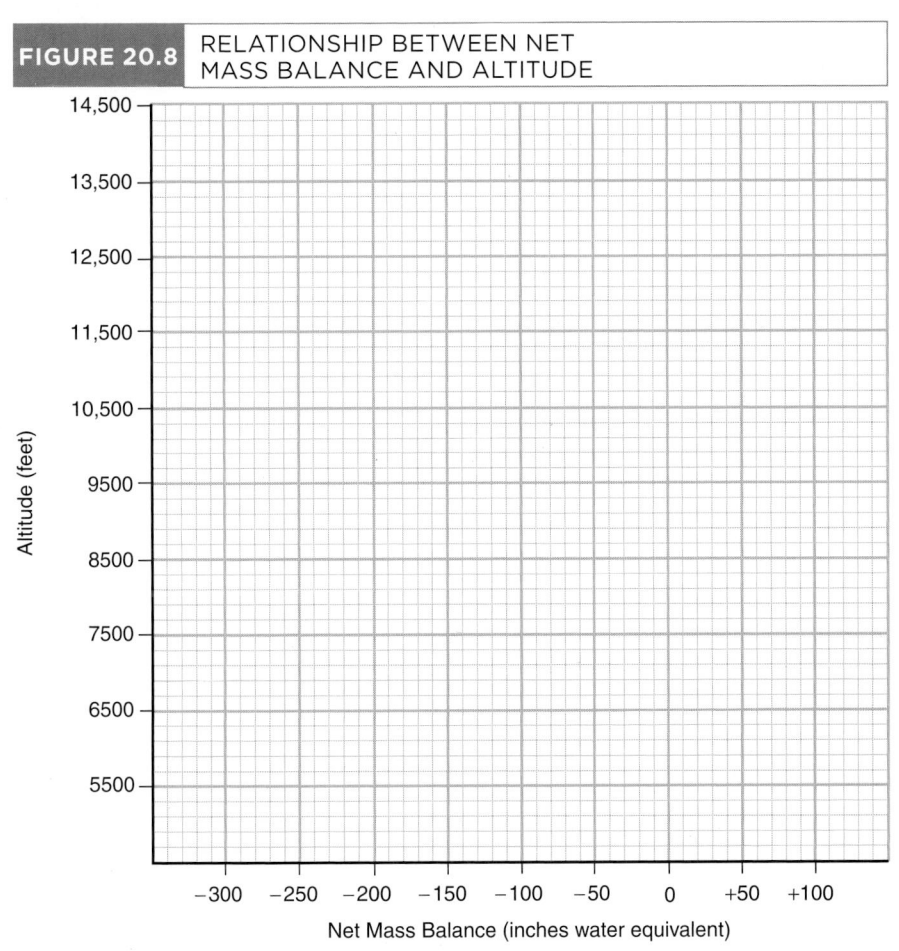

Name: _____ Section: _____

PART 2 † CONTINENTAL GLACIAL LANDFORMS

1. Examine Figure 20.9 showing the lobes of the Laurentide Ice Sheet and the Whitewater topographic map (Appendix E, Figure E.20).

   a. Based on these maps, what type of moraine would you call Kettle Moraine north-northeast of Palmyra?

   b. Based on these maps, where should we be able to find another example of this type of moraine (the same type as Kettle Moraine)?

   c. At its southern end, Kettle Moraine grades into another type of moraine that stretches to the west of Palmyra. How would we classify this type of moraine?

   d. All the area to the north and northwest of Kettle Moraine constitutes another type of moraine. What type of moraine is this?

   e. On the Whitewater topographic map (Appendix E, Figure E.20), label the regions where the moraines you listed appear. Note that not all of the moraines you listed appear on the topographic map.

2. a. A portion of Kettle Moraine is visible in the southeastern part of the Whitewater topographic map (Appendix E, Figure E.20). How does the map show the presence of the kettle holes that cover the moraine?

   b. In addition to finding kettle holes on moraines, where else in general might we expect to find them? Explain how kettle holes would form in this location.

3. a. If the Lake Michigan Lobe retreated from Wisconsin well before the Green Bay Lobe retreated at all (refer to Figure 20.9), what type of landform should we expect to find to the southeast of Kettle Moraine? Why?

   b. Examine the area to the southeast of Kettle Moraine on the Whitewater topographic map (Appendix E, Figure E.20) near Little Prairie and to the south of Little Prairie. Based on the appearance of the contour lines here, do you think this shows the landform you listed as an answer to part (a)? If so, explain why. If not, explain what it might be instead.

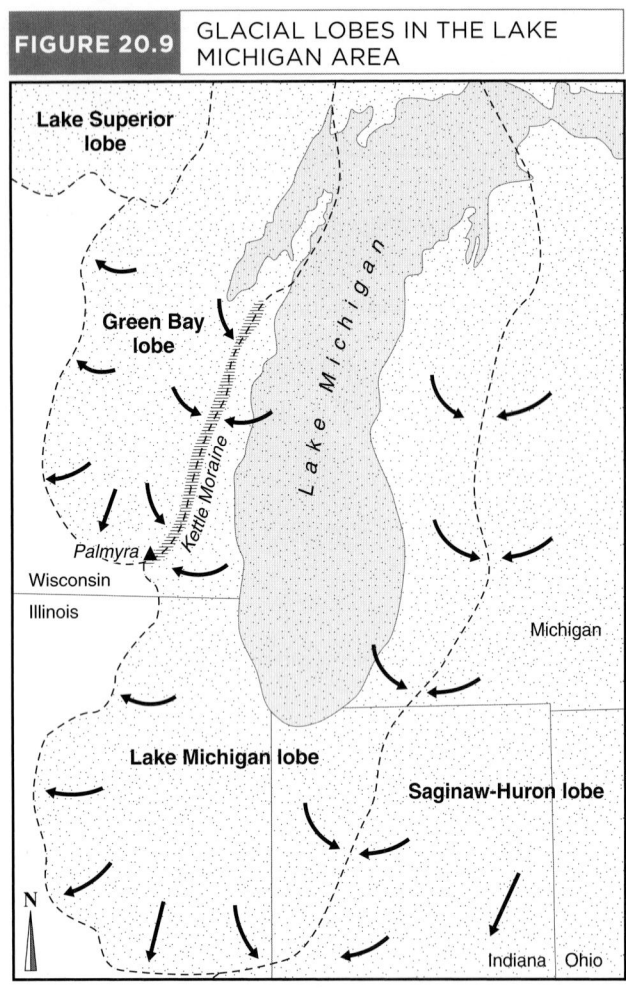

FIGURE 20.9 GLACIAL LOBES IN THE LAKE MICHIGAN AREA

4. Part of a drumlin field is evident in the northwestern portion of the Whitewater topographic map (Appendix E, Figure E.20). The Rome topographic map (Appendix E, Figure E.21) shows a larger scale version of these drumlins. Slabtown and Cushman Pond are two good reference points to help you match the maps.

   a. Select four drumlins on the Rome topographic map (Appendix E, Figure E.21). For each drumlin, select a contour line that approximates the base of the drumlin and highlight it. Record these base elevations below.

   b. Determine the highest elevation at the top of each of your four drumlins and record this information below.

   c. Determine the local relief for each of your four drumlins.

	Base Elevation (ft)	Top Elevation (ft)	Local Relief (ft)
Drumlin 1	_____	_____	_____
Drumlin 2	_____	_____	_____
Drumlin 3	_____	_____	_____
Drumlin 4	_____	_____	_____

   d. What is the average local relief for your drumlins? _____

e. Drumlins are elongated parallel to the direction of ice flow. Draw a line down the center of each of your drumlins showing the orientation of ice flow as indicated by the drumlin shape.

f. Examine the spacing of the contour lines at either end of the lines you drew for part (e). Based on the contour line spacing, and based on the lines you drew for part (e), what is the dominant ice flow direction for your drumlins?

g. Draw arrows on the Whitewater topographic map (Appendix E, Figure E.20) showing the ice flow direction.

EXERCISE 21 † COASTAL LANDFORMS

PURPOSE

The purpose of this exercise is to learn about various coastal processes and landforms.

LEARNING OBJECTIVES

By the end of this exercise you should be able to

- explain how wind and wave refraction affect wave direction and the direction of the longshore current;
- explain the processes forming and modifying coastal landforms;
- identify various coastal landforms on topographic maps and air photos;
- estimate rates of sea level change and explain why sea level change is not constant everywhere; and
- classify coasts as either emergent or submergent.

INTRODUCTION

Coastal areas are subject to processes related to wind, wave action, water currents, and sea level change. They are also areas where the characteristics of the biosphere, including humans, impact the effects of waves, currents, and wind on landforms. In this exercise, we will examine some of the impacts of wind, waves, water currents, and sea level change on the coastal landscape.

Wind and Waves

Wind is the major driving force behind ocean waves (although it is not the only force that generates waves). Wave size depends on three things: (1) wind speed, (2) wind duration, and (3) the **fetch**, or the length of open water over which the wind can blow in a particular direction. The largest waves are generated when the wind blows at high speeds in the same direction for a long time over a large area of water. These waves transmit large quantities of energy throughout the oceans, and this energy is transmitted to the continents along the coastal zone.

The actual direction a wave travels in is also affected by the fact that waves refract. **Refraction** refers to the bending of waves. As the ocean bottom interferes with the motion of water particles in waves, the waves slow down and bend. The result is that waves rarely move straight onshore; rather, they approach the shore at an angle. In addition, as waves strike headlands or other irregularities along the coast (including human-built structures such as breakwaters and jetties), they refract around these obstructions. The refraction of waves affects the spatial expenditure of energy along the coast, with energy being concentrated on headlands (Figure 21.1). Thus headlands are areas dominated by erosion while bays are dominated by deposition.

Currents

Because refraction causes waves to strike the coast at an angle, a **longshore current** is created. This longshore current ultimately results in sediment and water moving in a direction approximately parallel to the coast. The forward movement of water and sediment on shore at an angle is called **swash**. As waves recede, water and sediment travel straight back out to sea, creating **backwash**. This movement of swash and backwash results in particles following a zigzag pattern along the shore, creating a longshore current (Figure 21.2). The longshore current, similar to waves, transmits energy along the coast, eroding sediment from some locations and depositing it in others.

Coastal Landforms

The coastal zone is divided into three main areas—the nearshore zone, the foreshore zone, and the backshore zone (Figure 21.3). The **nearshore zone** starts at the seaward limit of wave action and extends to the line created by the lowest tides. The seaward limit of wave action is the location where the ocean bottom is shallow enough that it starts to interfere with movement of water in waves. The nearshore zone includes the area where waves break, creating surf. It also often contains sand bars and troughs that parallel the shore. Sand regularly moves back and forth between these offshore bars and the beach, with the beach growing and the bars shrinking during the summer season when swash dominates, and the beach shrinking and the bars growing during the winter season when backwash dominates.

The **foreshore zone** includes the intertidal area. It is the area repeatedly covered and uncovered by water as tides rise and fall. Tides follow a regular cycle, and in many locations, there are two high tides and two low tides per cycle. One of the high tides is generally higher than the other and one of

FIGURE 21.1 WAVE REFRACTION

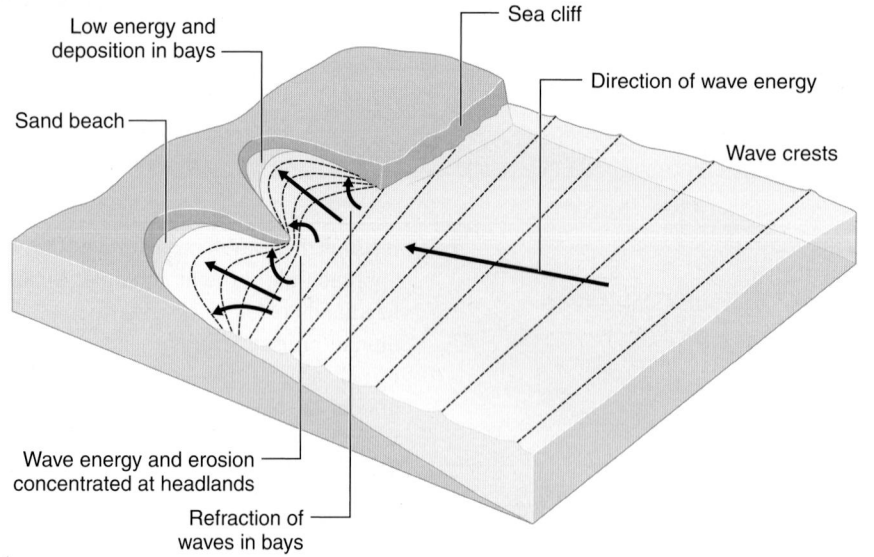

Wave refraction causes waves to bend. Refraction occurs when waves encounter shallower water causing waves to approach the shore at an angle. Refraction also occurs when waves strike headlands and bend into adjacent bays. Erosive energy is concentrated on the headlands, not in the bays.

FIGURE 21.2 LONGSHORE CURRENT AND DEPOSITIONAL LANDFORMS

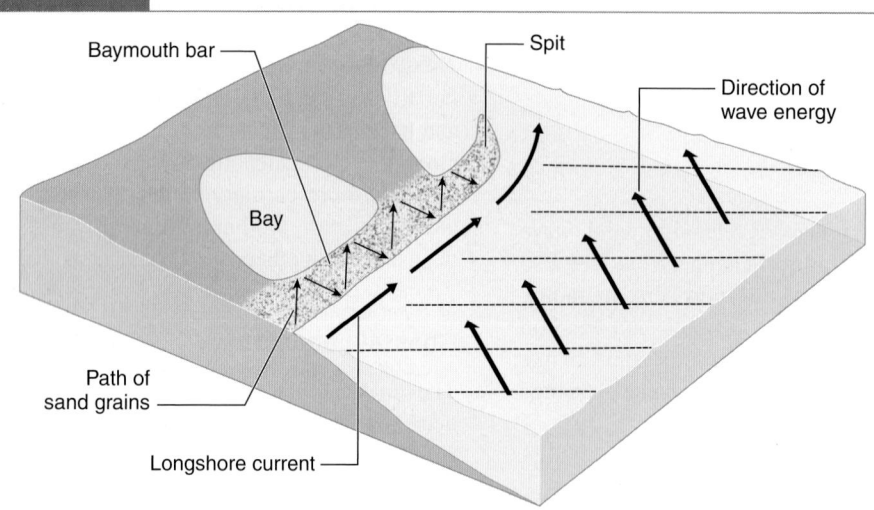

FIGURE 21.3 THE COASTAL ZONE

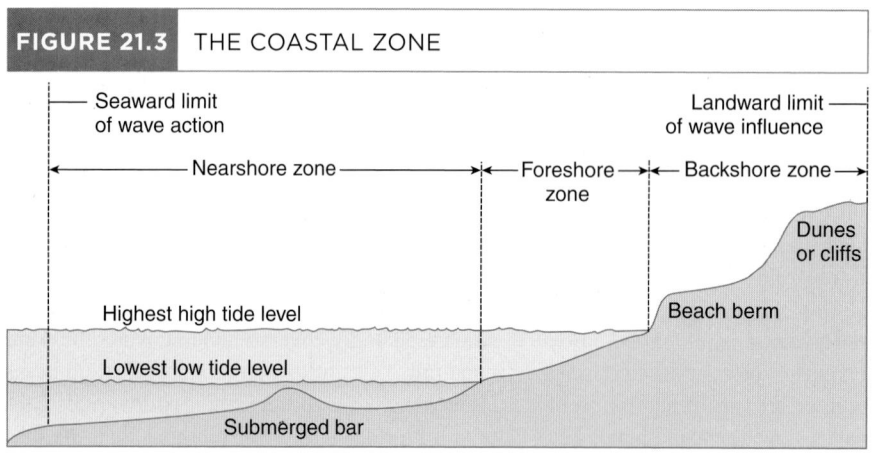

the low tides is generally lower than the other. Thus tidal records often refer to the *highest high tide* and the *lowest low tide*—the tidal extremes within one tidal cycle.

The **backshore zone** is the area landward of the highest high tide line and is often bounded by a beach berm; a relatively level sandy area. This area is generally dry except during storms when tides may run higher than usual and waves may be larger than usual. The landward edge of the backshore zone may be bounded by dunes, vegetation, or sea cliffs.

A variety of depositional landforms may form along beaches with a supply of sediment, usually sand (Figure 21.2). The longshore current will carry sediment along the coast until conditions change such that insufficient energy is available to carry the sediment farther, at which time the sediment is deposited. This often happens when the longshore current encounters a bay. The result is the formation of a **spit,** a narrow strip of sand connected to the shore. The spit grows in the direction of the longshore current. Continued growth of a spit may result in the formation of a **baymouth bar,** which can block the bay from the direct impact of waves. In coastal areas with rocky headlands bordered by bays, waves tend to lose their energy upon entering the bay and these bays may also become areas of sand accumulation, forming sandy beaches.

Narrow strips of sand that are parallel to and not connected to the shore are called **barrier islands.** Barrier islands have sandy beaches on their coastal side, a region of sand dunes, and salt marshes on their back side. A lagoon often separates the barrier island from the mainland. Barrier islands bear the brunt of wave energy, protecting the mainland coast from the direct impact of tides and storm waves. As sea level rises, waves may wash over the barrier island, moving sand from the coastal side to the landward side, causing the island to migrate landward.

The local sediment budget plays an important role in the formation, modification, and destruction of depositional landforms. A sediment budget is an accounting of the total amount of sand added to, and removed from, a particular stretch of shoreline over some period of time (such as a year or ten years). Sediment is contributed to coastal areas by rivers, by the longshore current, and by wind and waves. The longshore current, wind, and waves can also remove sediment from the coast. Since rivers, currents, waves, and wind are always at work, the shoreline is a very dynamic environment, with sediment constantly being added and removed simultaneously. So, although to us the coast may appear constant from one visit to the next, sediment is always on the move. Shorelines with a positive sediment budget will experience growth of beaches, sand bars, spits, and barrier islands. Growth may result in beaches, barrier islands, and sand bars widening and/or growing taller, or spits growing longer. A negative sediment budget will have the opposite effect.

In areas without a supply of sand or a negative sediment budget, and particularly in areas with rocky coasts, we can find a variety of erosional landforms (Figure 21.4). Emergent rocky coasts are often characterized by **sea cliffs.** As waves pound against the base of the cliff, a **wave-cut notch** may slowly form at the base of the cliff. Over time, the notch may grow to form a cave. Eventually, the overlying rock may collapse. As rocky cliffs are eroded away, **sea stacks** may form, tall isolated spires of rock indicating the former location of the shore. The rocky area over which the waves travel may also be abraded to a relatively smooth surface by the sand and sediment carried along with the waves. This relatively smooth surface is called a **wave-cut platform.** In tectonically active areas, the combined effects of changes in sea level and tectonic uplift may produce a series of wave-cut platforms, or a series of "steps," called **marine terraces.**

Wind and Sand

In sandy coastal areas, waves and currents are not the only forces shaping the coastline. Wind also impacts the accumulation of sand and the appearance of the coast. The combined action of waves and wind may result in the formation of a variety of dunes. Vegetation helps to stabilize these dunes. The most common type of dune is a **foredune,** which forms a ridge at the back of the beach above and parallel to the high tide line. When the foredune is vegetated and stable, it may form a continuous ridge; but in areas of sparse vegetation, it may appear as a series of vegetated mounds, rather than a continuous ridge. If the sediment budget along the beach is positive and the beach builds seaward, a series of parallel dunes may form. The seaward-most dune is called the foredune and the older landward dunes are called *secondary dunes.* The ongoing movement

FIGURE 21.4 EROSIONAL LANDFORMS

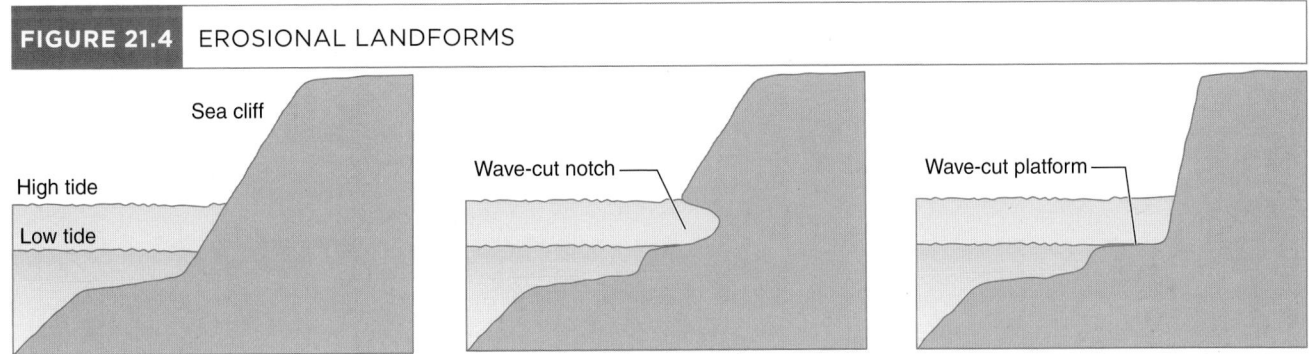

of sand by the wind may result in some other types of dunes. In areas where the vegetation has been disturbed, blowouts and parabolic dunes may form. **Parabolic dunes** are U-shaped with their "arms" facing into the wind.

Changes in Sea Level

The impact of waves on the coast is affected by the water level—higher water levels impact locations farther inland than lower water levels. As a result, changes in sea level have greatly impacted the coastal landscape. **Transgression** refers to rising sea levels; **regression** refers to falling sea levels.

Changes in sea level result from two major processes, eustatic changes and isostatic changes. **Eustatic changes** are changes in sea level due to changes in the quantity and temperature of water in the oceans. These changes are often associated with the progression of glacials and interglacials. During glacials, water is removed from the oceans and added to ice sheets, lowering sea level. In addition, as the oceans cool, the water contracts somewhat, which further adds to the decline in sea level (although if water cools enough to freeze, it expands). During interglacials, as climatic conditions warm up, glaciers melt and water is returned to the oceans, causing sea level to rise. Presently, many glaciers throughout the world are retreating and are contributing increasing amounts of water to the world's oceans. In addition, as the climate of the earth warms, the ocean water warms and expands somewhat (similar to how air expands when it warms). The result is that we are currently experiencing a period of sea level rise.

The second process affecting sea level relates to **isostatic changes.** These are changes in the level of the land surface due to changes in the weight of the continents or due to tectonic activity. For example, as rivers contribute sediment to the coastal zone, this sediment accumulates and adds weight to the land surface, which can cause subsidence. To us, this would look like an increase in sea level. In tectonically active areas, faulting may result in the land surface being uplifted. To us, this might appear as a decrease in sea level.

These two processes working independently or in conjunction result in an **apparent change** in sea level, which is a concern to humans given that a large percentage of the world's population lives near the coast, and given that many people living inland like to vacation along the coast. As sea level changes, this impacts not only the coastal landscape but the human landscape and human activities as well.

Coastlines that are experiencing an apparent drop in sea level are called **emergent coastlines.** These coastlines are characterized by sea cliffs, wave-cut platforms, and marine terraces. Coastlines that are experiencing an apparent increase in sea level are called **submergent coastlines.** These coastlines are characterized by submerged river valleys, tidal marshes, and in areas of abundant sand, extensive sandy beaches. Thus these two different types of coasts often have different landscape characteristics.

IMPORTANT TERMS, PHRASES, AND CONCEPTS

fetch	sea cliff
refraction	wave-cut notch
longshore current	eustatic change
swash	isostatic change
backwash	apparent sea level change
transgression	emergent coastlines
regression	submergent coastlines
nearshore zone	sea stack
foreshore zone	wave-cut platform
backshore zone	marine terrace
spit	foredune
baymouth bar	parabolic dune
barrier island	

COASTAL LANDFORMS

1. Figure 21.5 shows the location of two data collection buoys (WIS44013 and WIS44005) off the coast of Cape Cod, and a third buoy (WIS57) for which wind and wave data have been generated from numerical models by the U.S. Army Corps of Engineers. The charts in Figure 21.6 show the percent of time the wind or waves come from a particular direction for these three buoys.

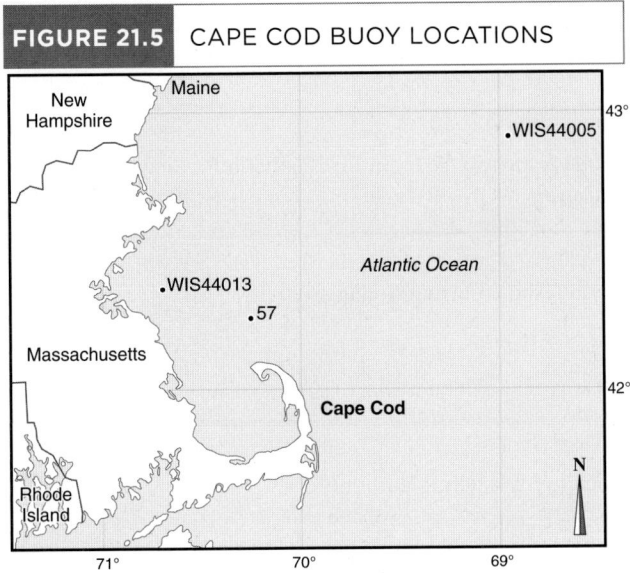

FIGURE 21.5 CAPE COD BUOY LOCATIONS

Source: U.S. Army Corps of Engineers.

a. From what direction is the wind and are waves most likely to come for each buoy?

	Wind	**Waves**
WIS44005	_____	_____
WIS44013	_____	_____
WIS57	_____	_____

b. Draw three red arrows on Figure 21.5, one for each buoy, showing the prevailing wave direction for that buoy. Draw three more arrows in blue, one for each buoy, showing the prevailing wind direction for that buoy. Draw all of these arrows from the edge of the map to the appropriate buoy.

c. Based on the wind patterns alone, what direction would you expect waves to come from for each buoy?

d. Do the wind direction arrows (blue) you drew for each buoy primarily cross water or land?

e. Which buoy's prevailing wave direction best matches its prevailing wind direction?

f. Why do you think waves for the other two buoys don't match the buoys' wind directions very well?

275

g. Explain the process responsible for the prevailing wave direction at these two buoys.

h. Based on the wind and wave data, in what direction would you infer the longshore current is flowing on the north side of Cape Cod? Why?

2. a. What clue do you see on the topographic map of Cape Cod (Appendix E, Figure E.22) that might indicate the direction of the longshore current on the north or west side of the Cape?

b. Based on your answer to part (a), in what direction is the longshore current flowing?

c. Is this the same direction you inferred from the wind and wave data in question 1 (h)?

d. With this longshore current, what could happen to the opening to Hatches Harbor?

3. a. Examine the air photo of Cape Cod (Figure 21.7). On the air photo, draw lines along several wave crests off the north shore and off of Race Point.

b. Based on the pattern of the wave crests on the air photo, in what direction is the longshore current flowing?

c. How has the entrance to Hatches Harbor changed from 1972 (the date of the map) to 1995 (the date of the air photo)?

d. Where on the air photo do you see a potentially new spit forming?

4. Locate the offshore bar named Peaked Hill Bar on the Cape Cod topographic map (Appendix E, Figure E.22).

a. Approximately how deep is the water over the highest point on this bar? Remember that bathymetric contours show water depths and do not follow the contour interval.

b. Approximately how many miles offshore is the top of this bar?

FIGURE 21.6 — CAPE COD WIND AND WAVE ROSES

WIS 44013 Wind direction (percent of time)

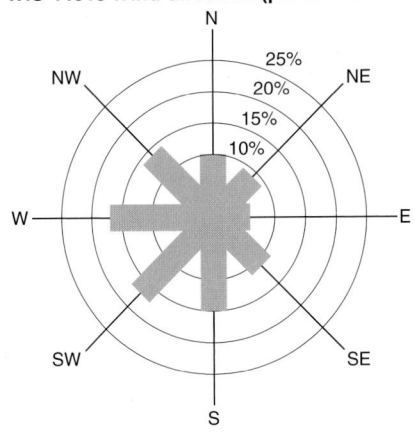

WIS 44013 Wave direction (percent of time)

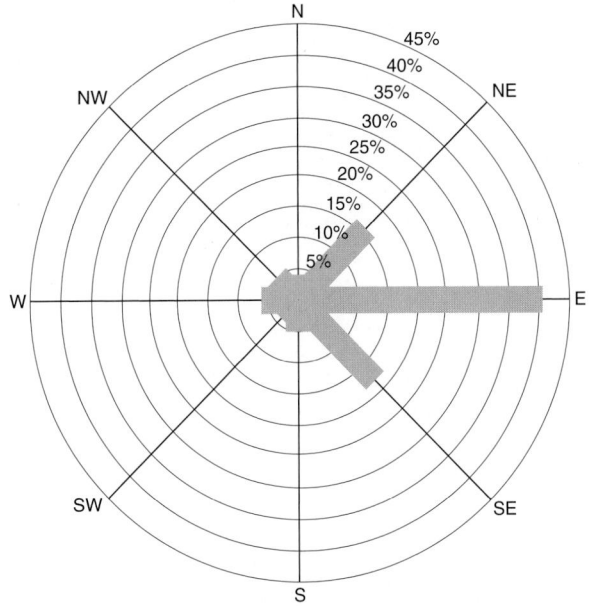

WIS 57 Wind direction (percent of time)

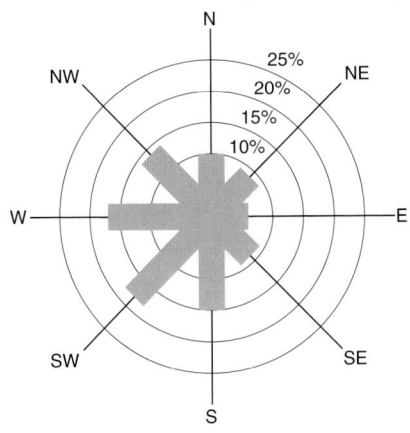

WIS 57 Wave direction (percent of time)

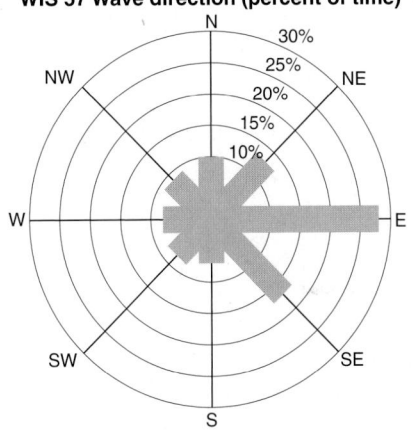

WIS 44005 Wind direction (percent of time)

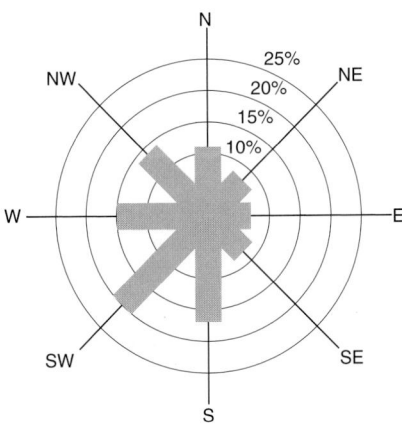

WIS 44005 Wave direction (percent of time)

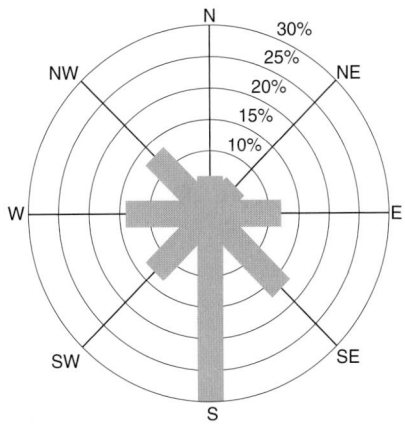

Bars show the direction the wind or waves are *coming from*; thus a bar stretching to the east indicates wind or waves coming from the east. The longer the bar, the greater the percent of time the wind or waves come from that direction. For example, at station 44013, winds come from the W and SW more often than from any other direction (about 17% of the time), while waves come from the east about 40% of the time and from the west only about 6% of the time.

277

| FIGURE 21.7 | AIR PHOTO OF CAPE COD, 1995 |

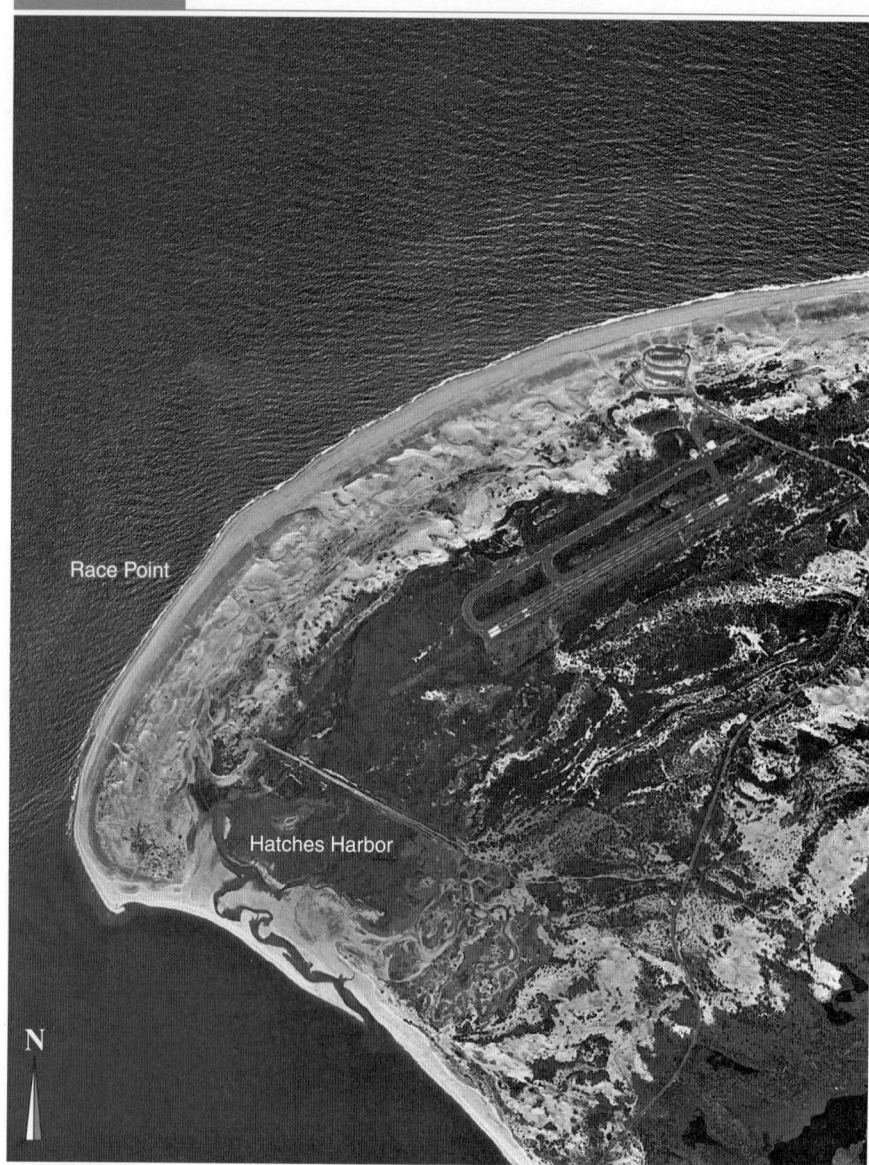

Source: Courtesy of the USGS.

c. Explain how sand is exchanged between offshore bars and beaches.

d. Based on your answer to part (c), predict how the size of the Peaked Hill Bar, as reflected by the water depth, will change from summer to winter.

278

5. Some large sand dunes are evident on the Cape Cod topographic map (Appendix E, Figure E.22) north of Pilgrim Lake.

   a. Outline three or four of these dunes.

   b. What direction are the winds that created these dunes coming from?

   c. Figure 21.8 contains data on wind speed from Station WIS57. Examine these data along with the wind direction data in Figure 21.6. What seems to be more important in shaping the dunes, the direction the wind comes from most often or the direction the strongest winds come from?

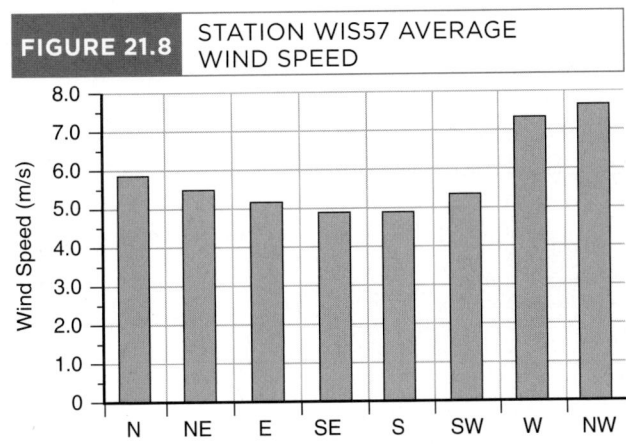

FIGURE 21.8 STATION WIS57 AVERAGE WIND SPEED

   d. Based on the wind patterns and the shape of these dunes, what type of dunes do you think they are? Why?

6. Figure 21.9 shows the location of Fire Island National Seashore. Appendix E, Figure E.23 shows a portion of the topographic map of Fire Island National Seashore, and Figure 21.10 shows a topographic profile across the island.

   a. What kind of landform is Fire Island?

   b. Label the foredune and the secondary dune on Figure 21.10 and on the topographic map (Appendix E, Figure E.23).

   c. Mark the edge of the beach berm on the profile (Figure 21.10) and on the topographic map (Appendix E, Figure E.23).

   d. What does the existence of a secondary dune indicate about the sediment budget at this location? Explain why.

   e. Based on the topographic map, does the foredune appear as a continuous ridge or as a series of sand mounds?

   f. What does this indicate about the presence of vegetation along this portion of Fire Island?

FIGURE 21.9 FIRE ISLAND LOCATION MAP

Source: U.S. National Park Service.

FIGURE 21.10 NORTH-SOUTH PROFILE ACROSS FIRE ISLAND AT THE SUNKEN FOREST. VERTICAL EXAGGERATION IS APPROXIMATELY 10X

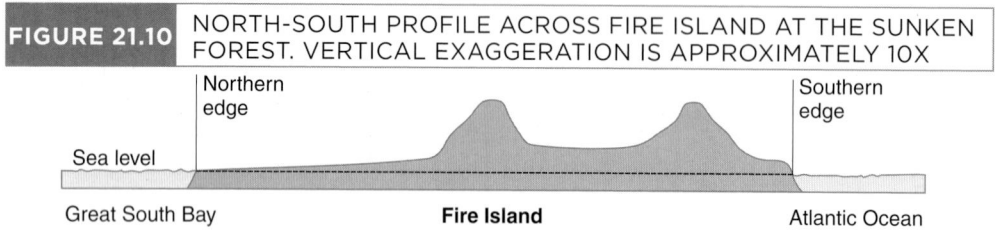

7. a. How wide is Fire Island at the widest point of the Sunken Forest, in feet? Use the topographic map (Appendix E, Figure E.23) to measure this.

b. Scientists estimated the average shoreline landward displacement of Fire Island was 0.4 m/yr (1.3 ft/yr) from 1870–1998. At this rate, calculate how many years it would take for the island at the widest point of the Sunken Forest to move an island's width landward. Show your work.

8. Data on apparent sea level rise have been collected at stations along the east coast of the United States. Data from Sandy Hook, New Jersey, which is geomorphologically similar to Fire Island, are shown in Figures 21.11 and 21.12. *Mean annual sea level* represents the average sea level over the course of a year. *Mean annual highest high tide* is the average of the highest high tides throughout the year. The tidal cycle here shows two high tides per cycle, and these data are based on the higher of these two high tides.

a. Use a ruler to draw a straight trend line through the center of the points on each graph. Your trend lines should extend from the year 1930 to the year 2020.

b. Based on your trend lines, determine the following elevations:

	Mean Annual Sea Level	Mean Annual Highest High Tide
2020	_____	_____
1930	_____	_____

| FIGURE 21.11 | MEAN ANNUAL SEA LEVEL, SANDY HOOK, NEW JERSEY |

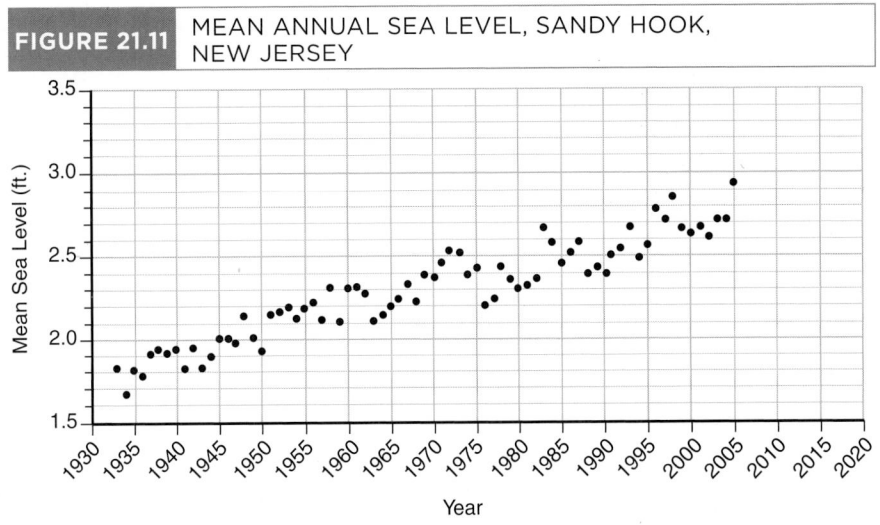

Source: Data from NOAA.

| FIGURE 21.12 | MEAN ANNUAL HIGHEST HIGH TIDE, SANDY HOOK, NEW JERSEY |

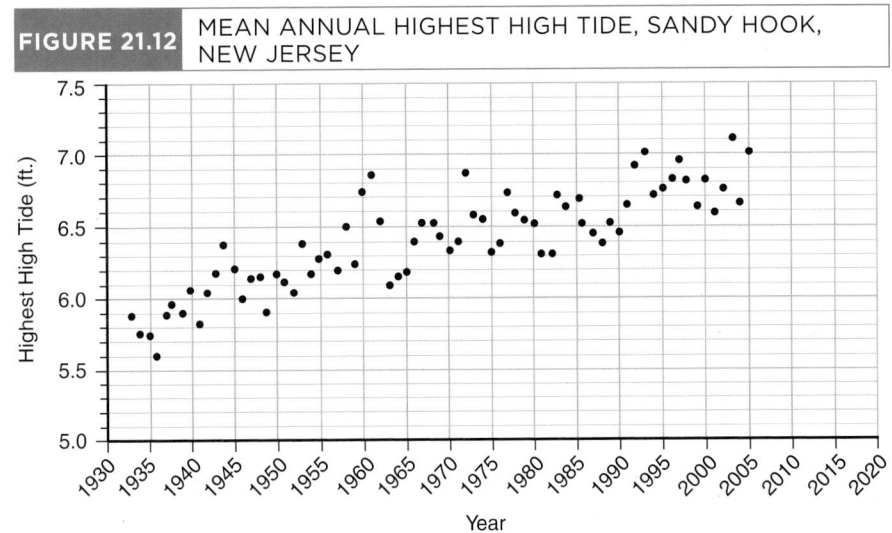

Source: Data from NOAA.

c. Use the information in part (b) to calculate the rate at which sea level has risen per year for each of these measurements. Show your work.

Annual increase in mean annual sea level:

Annual increase in mean annual highest high tide:

d. How many feet above sea level is the highest point in the Sunken Forest?

e. If sea level continues to rise at the rate you calculated in part (c), how many years will it take for the mean annual highest high tide to overtop the highest elevation in the Sunken Forest, assuming the average highest high tide in 2020 is 7.2 feet?

f. What elevations occur in the built-up areas?

g. How many years will it take for the mean annual highest high tide to invade the built-up areas, using the rate you calculated in part (c) and assuming the average highest high tide in 2020 is 7.2 feet?

h. As sea level rises over the next 30 to 50 years, storm tides and waves will get progressively higher. Storm tides and waves are not accounted for in the annual data shown in Figures 21.11 and 21.12, thus it is possible that storm tides and waves may overtop the highest elevations sooner than you predicted in parts (e) and (g). As this happens, how do you think the island will change? Do you think landward migration will accelerate or might the island possibly break apart into discontinuous segments? Why?

9. Examine the topographic map of Palos Verdes, California, (Appendix E, Figure E.24) and the associated profile in Figure 21.13.

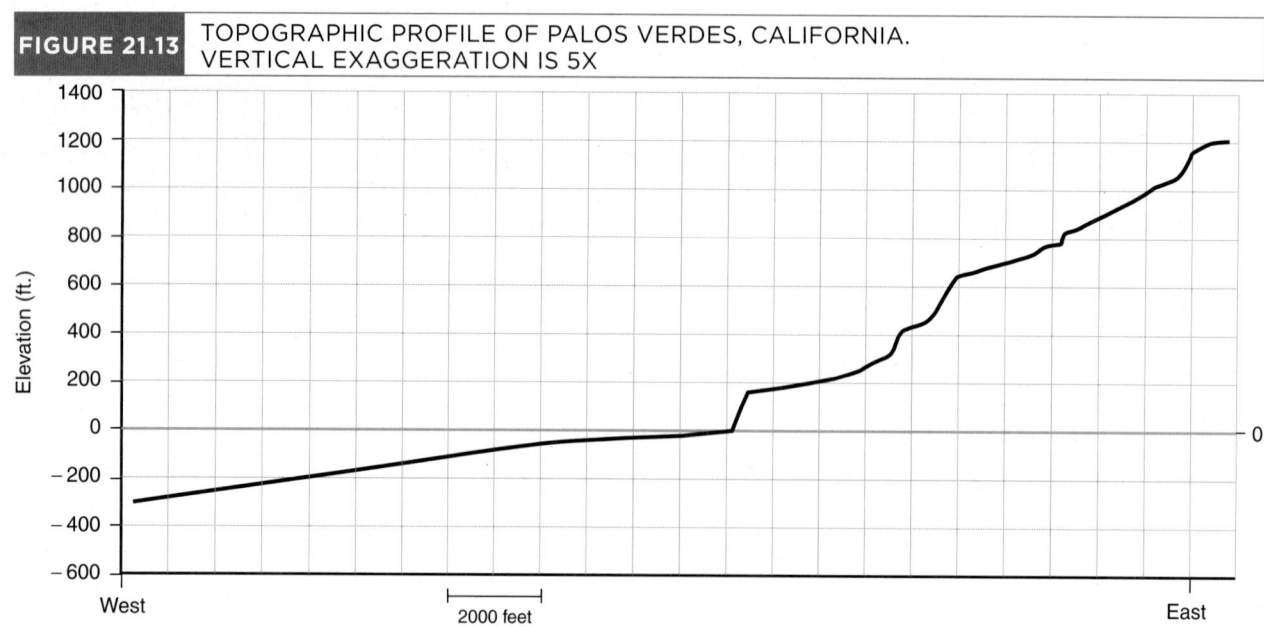

FIGURE 21.13 TOPOGRAPHIC PROFILE OF PALOS VERDES, CALIFORNIA. VERTICAL EXAGGERATION IS 5X

Note: Profile scale is smaller than that of the corresponding topographic map in Figure E.24.

a. What might be an explanation for why the offshore (underwater) gradient is so flat adjacent to the shoreline?

b. What kind of landform could this relatively flat area be an example of?

c. Mark any marine terraces on the profile and on the map.

d. How many episodes of uplift are indicated by the marine terraces at this location?

10. Examine the air photo (Figure 21.14) of Palos Verdes.

FIGURE 21.14   PALOS VERDES AIR PHOTO, 1994

Source: Courtesy of the USGS.

a. On the air photo, draw lines along several wave crests that have collided with Resort Point and entered the bays on either side of the point.

b. What happens to the waves as they pass Resort Point and enter the bays on either side of it?

c. Draw lines perpendicular to the wave crests ending in arrows to show where wave energy will be concentrated (these will not be straight lines). See Figure 21.1 as an example.

d. Mark with a star the location on the air photo where the main erosive energy of these waves will be focused.

e. Given sufficient time, what could happen to Resort Point?

f. Identify some examples of sea stacks on the air photo (Figure 21.14) and on the topographic map (Appendix E, Figure E.24), and circle them.

g. In general, where do sea stacks seem most likely to form?

11. Data on apparent sea level rise collected at Los Angeles, California, of which Palos Verdes is a southern suburb, are shown in Figure 21.15.

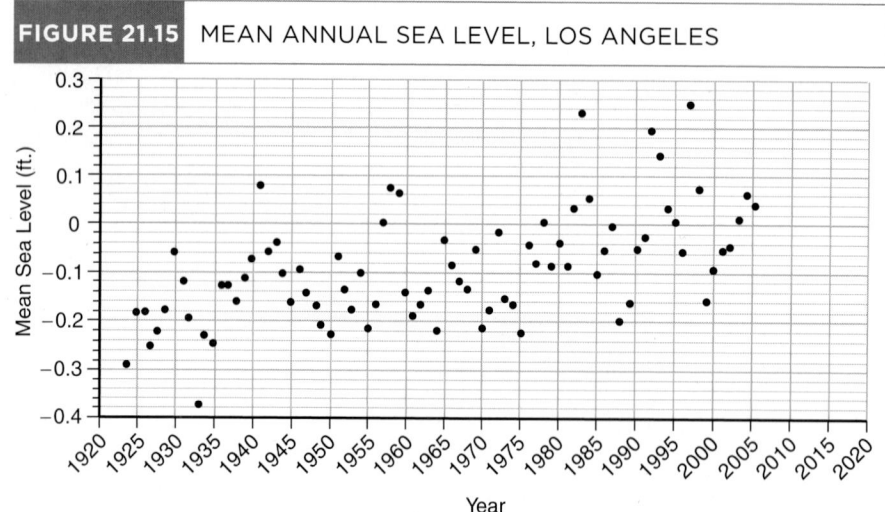

FIGURE 21.15 MEAN ANNUAL SEA LEVEL, LOS ANGELES

Source: Data from NOAA.

a. Use a ruler to draw a straight trend line through the center of the points on the graph. Your trend line should extend from the year 1920 to the year 2020.

b. Based on your trend line, determine the following elevations:

   Mean Annual Sea Level

   2020 _____

   1920 _____

c. Use the information in part (b) to calculate the rate at which sea level has risen per year. Show your work.

   Annual increase in mean annual sea level:

d. How does the annual rate of sea level change on the west coast compare to the annual rate of sea level change on the east coast (question 8 (c))?

e. What might account for the difference in the annual rate of sea level change along the east and west coasts of the United States? (Hint: Think about whether eustatic or isostatic changes will prevail on the east coast or the west coast.)

12. a. Coasts may be classified as either submergent or emergent based on changes in sea level and based on the landforms found along the coast. Considering these two pieces of information independently, how would you classify the three locations explored in this exercise?

	Classification based on Sea Level Change	**Classification based on Landforms**
Cape Cod, Mass.	_____	_____
Fire Island, N.Y.	_____	_____
Palos Verdes, Calif.	_____	_____

b. If your classification based on sea level change doesn't match your classification based on landforms for any one of these locations, what might be an explanation for why your answers are different?

APPENDIX A
UNITS OF MEASURE AND CONVERSIONS

TABLE A.1 LENGTH

1 km = 100,000 cm	1 m = 0.001 km	1 cm = 0.01 m	1 mm = 0.1 cm
= 1000 m	= 100 cm	= 10 mm	= 0.001 m
= 3281 ft	= 39.4 in	= 0.0328 ft	= 0.0394 in
= 0.621 mi	= 3.281 ft	= 0.394 in	
1 mi = 5280 ft	1 ft = 12 in	1 in = 25.4 mm	
= 63,360 in	= 30.48 cm	= 2.54 cm	
= 1.61 km	= 0.3048 m		
= 1609 m			

TABLE A.2 AREA

1 km^2 = 0.386 mi^2	1 m^2 = 10,000 cm^2	1 cm^2 = 100 mm^2	1 mm^2 = 0.01 cm^2
= 100 ha (hectares)	= 1550 in^2	= 0.155 in^2	= 0.00155 in^2
	= 10.76 ft^2		
1 mi^2 = 27,878,400 ft^2	1 ft^2 = 144 in^2	1 in^2 = 6.452 cm^2	1 ha = 10,000 m^2
= 2.590 km^2	= 929 cm^2		
= 640 ac (acres)			
= 259 ha (hectares)			

TABLE A.3 VOLUME

1 km^3 = 10^9 m^3	1 m^3 = 10^6 cm^3	1 cm^3 = 0.061 in^3	1 ml = 0.016 in^3
= 0.24 mi^3	= 35.31 ft^3	= 1 ml	= 1 cm^3
	= 264.2 gal		
	= 1000 l		
1 ft^3 = 1728 in^3	1 in^3 = 16.39 cm^3	1 gal = 3.79 l	1 l = 61.02 in^3
= 7.48 gal			= 0.2642 gal
= 28,317 cm^3			= 1000 ml
= 28.32 l			

TABLE A.4 WEIGHT

1 kg = 1000 g
     = 2.205 lb
	1 g = 0.001 kg
	    = 0.0353 oz
	    = 0.0022 lb
		1 lb = 16 oz
		     = 0.4536 kg
		     = 453.6 g
			1 oz = 28.35 g

1 US ton = 2000 lb
         = 907.2 kg
	1 metric ton = 106 g
	            = 2205 lb
	            = 1.1 U.S. tons		

TABLE A.5 PRESSURE

1 bar = 14.5 lb/in²
      = 100,000 Pa
	1 mb = 0.001 bars
	     = 100 Pa
	     = 1 hPa
	     = 0.0295 in Hg
		1 atmosphere = 1013.25 mb
		             = 101,325 Pa
		             = 1013.25 hPa
		             = 29.89 in Hg

mb = millibars	Pa = pascals	hPa = hectopascals	Hg = mercury
1 atmosphere = air pressure at sea level			

TABLE A.6 VELOCITY

1 m/sec = 3.6 km/hr
       = 2.237 mi/hr
	1 km/hr = 0.2278 m/sec
	        = 0.9113 ft/sec
		1 ft/sec = 0.6818 mi/hr
		         = 1.097 km/hr
			1 mi/hr = 1.4767 ft/sec
			        = 0.447 m/sec

TABLE A.7 TEMPERATURE

°F = 1.8 °C + 32

°C = (°F − 32)/1.8

K = °C + 273.15

TABLE A.8 ENERGY

1 cal = 4.1868 J	1 W = 1 J/sec
	    = 0.239 cal/sec
	    = 14.34 cal/min
	1 W/m² = 0.001433 ly/min
		1 cal/sec = 4.1868 W

cal = calorie	W = Watts	J = Joule	ly = Langley = 1 cal/cm²

TABLE A.9 TANGENT OF AN ANGLE (ANGLES IN DEGREES)

Angle	Tangent	Angle	Tangent	Angle	Tangent	Angle	Tangent	Angle	Tangent
1°	0.0175	10°	0.1763	19°	0.3443	28°	0.5317	37°	0.7536
2°	0.0349	11°	0.1944	20°	0.3640	29°	0.5543	38°	0.7813
3°	0.0524	12°	0.2126	21°	0.3839	30°	0.5774	39°	0.8098
4°	0.0699	13°	0.2309	22°	0.4040	31°	0.6009	40°	0.8391
5°	0.0875	14°	0.2493	23°	0.4245	32°	0.6249	41°	0.8693
6°	0.1051	15°	0.2679	24°	0.4452	33°	0.6494	42°	0.9004
7°	0.1228	16°	0.2867	25°	0.4663	34°	0.6745	43°	0.9325
8°	0.1405	17°	0.3057	26°	0.4877	35°	0.7002	44°	0.9657
9°	0.1584	18°	0.3249	27°	0.5095	36°	0.7265	45°	1.0000

APPENDIX B
DRAWING ISOLINES

One way to show the geographic distribution of phenomena is to construct a map using isolines. **Isolines** are lines that connect points of equal value. The values might be annual amounts of precipitation, average annual temperatures, or elevation. Maps that use isolines are called *isarithmic maps*. An important reason for constructing an isarithmic map is to make what might appear to be a hodge-podge distribution of point data easier to use and interpret. Figure B.1 shows what at first glance appears to be a random set of points distributed about the map. In this example, the data points represent total annual rainfall. Isolines connecting points with equal precipitation amounts are called *isohyets*.

Close inspection shows that the precipitation values tend to increase from the lower right corner to the upper left corner of the map. Before drawing any isolines, decide on an appropriate interval to separate each isoline (the difference in amount between successive isolines). The interval determines the number of isolines to draw on the map. It would be foolish to have an isoline for each value (an interval of 1) because the map would become too cluttered with lines. Too large of an interval would yield few isolines and the distributional pattern might not be discernable. An interval of 4 seems to be an appropriate interval, given the range of values in the data for Figure B.1A. Isolines with whole values are used as it makes interpolation of values between isolines, where no data exist, easier.

Given an interval of 4 units and the precipitation values plotted on the map, the first line drawn will have a value of 12. Starting in the lower left corner of Figure B.1B, the 12-inch isoline will connect the two data points having a value of 12. But where will the line go from there without any other values of 12 to guide us? Interpolation of where other values of 12 are located can be accomplished by using other nearby data points. Figure B.1C illustrates how to draw an isoline using interpolation.

Notice in Figure B.1C the position of a 16-inch isoline was drawn using interpolation. This was done first by evenly dividing the space between the 14 and 18 on the left side of the map. The value of 16 was found and the line was started from there. The same thing was done between the values of 15 and 20 on the right side of Figure B.1C. The value of 16 was found and the 16-inch isoline was drawn across the map.

Returning to Figure B.1B, the 12-inch isoline is drawn between the 11- and 14-inch values by interpolating its position in the same way as the example in Figure B.1C. Before going on, check to see if the position of the line is correct. Values less than 12 should be below the line and those larger than 12 above the line. When this has been verified, proceed with the next isoline. Given that the interval is 4, the next line to draw is the 16-inch isohyet. Once the 16-inch isohyet is drawn, check its position. Values less than 16 but greater than 12 should be located between the 16- and 12-inch isohyets. Values greater than 16 should fall above the line. Last, the 20-inch isohyet is drawn.

The spacing of isolines illustrates how much change in the mapped property there is over distance or, a *gradient*. The closer the spacing, the more rapid the change over distance. Figure B.2 gives two examples of isoline spacing. In this case, the isolines are *contour lines*, lines that connect points of equal elevation. In either Figure B.2A or B.2B, notice that the isolines cover the same distance. In the first instance (B.2A), the contour lines are closely spaced between one another. This would indicate that elevation is changing rapidly over a short distance, i.e., a steep gradient. The isolines in Figure B.2B

FIGURE B.1 DRAWING AND INTERPOLATING ISOLINES

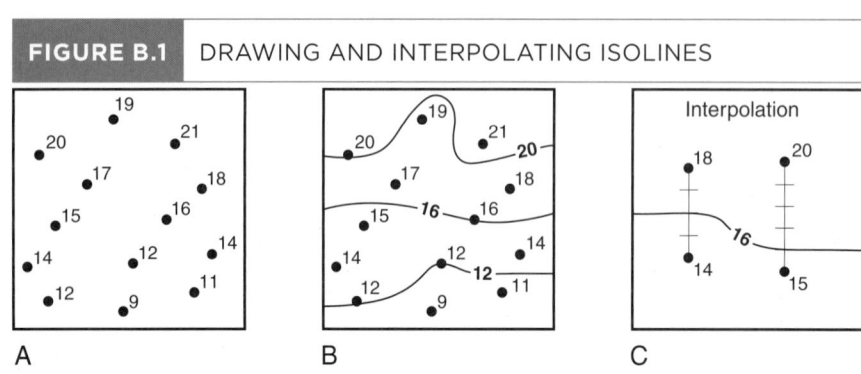

A B C

290

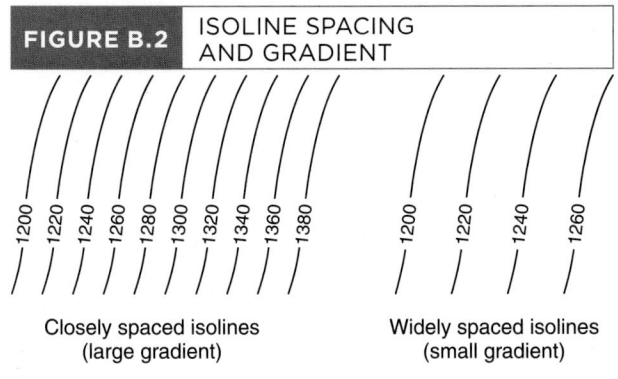

FIGURE B.2 ISOLINE SPACING AND GRADIENT

Closely spaced isolines (large gradient)

Widely spaced isolines (small gradient)

are more widely spaced but cover the same distance. This means that less change in elevation occurred, so there must be a gentle slope or small gradient.

Isolines may also be enclosed. *Enclosed isolines* indicate that values are higher inside the enclosure than outside, or vice versa. Figure B.3 shows the distribution of pressure across the United States. Isolines connecting points of equal air pressure are called *isobars*. Notice that a well-developed area of low pressure, indicated by the enclosed isobars, is found along the east coast in southern New Jersey, Delaware, and Maryland. An area of high pressure, also indicated by closed isobars, is located in western Arizona.

Even though isolines may be enclosed, isolines with different value should never cross. This would mean that a single point would have two different values.

FIGURE B.3 PRESSURE MAP OF THE UNITED STATES

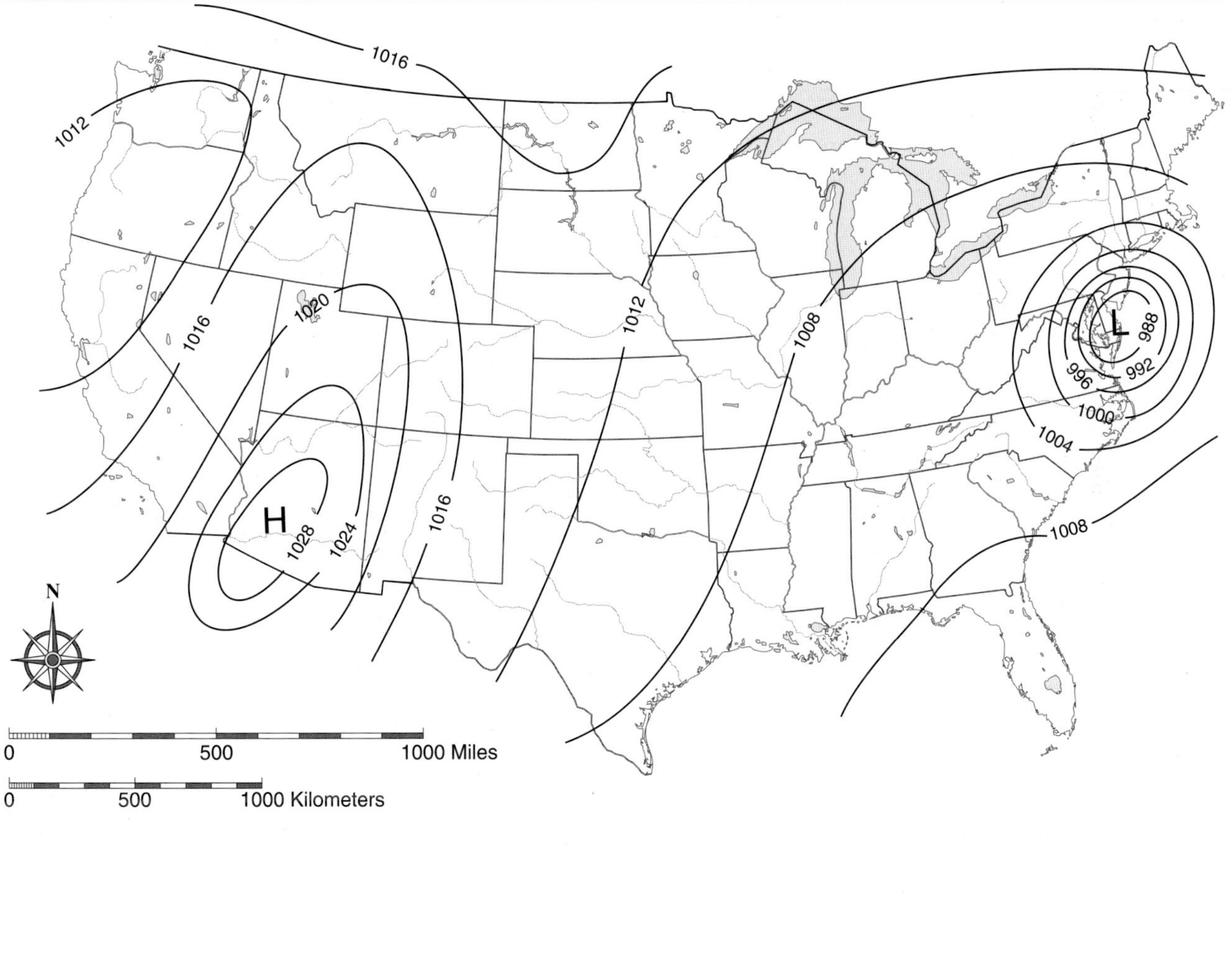

APPENDIX C
CONSTRUCTING PROFILES

Profiles are used in conjunction with isoline maps. Profiles provide a "picture" of whatever feature an isoline map is displaying. A profile drawn from a topographic map provides a side-view picture of the shape of the landscape. A profile drawn from an isobar map provides a visual image of how atmospheric pressure varies across space. Whether an isoline map shows something we can actually "see," such as topography, or whether it shows something we cannot actually "see," such as atmospheric pressure, profiles help us visualize how that particular feature changes over space.

The first step for drawing a profile is to pick two points on the map, between which you want to know how some feature changes, such as topography. Draw a line on the map connecting the two end points you have chosen, and label the end points. The end points are usually labeled with capital letters.

Next, lay the edge of a separate piece of paper along the line drawn on the map. On that piece of paper, mark and label the location of the two end points. Then, wherever an isoline intersects the piece of paper, make a tick mark on the paper and note the value associated with that isoline. These tick marks will be spaced along the edge of the paper exactly the same as on the map.

Third, carefully transfer these tick marks to the bottom of a piece of graph paper along with their numerical values. The horizontal axis along which the tick marks are placed represents distance across the landscape and has a scale identical to the scale of the map from which the information was taken.

The fourth step involves setting up the vertical axis for the graph. Find the lowest and highest values along the profile line. These values indicate the range of values that must be included on the vertical axis of the graph. The vertical axis does not need to start at zero; it may start at the lowest isoline value, or some convenient number less than that. The vertical axis may end at the highest isoline value, or some convenient number greater than that. For example, if the range of elevations for a topographic profile is 273 feet to 1947 feet, the vertical axis might range from 250 to 2000 feet. Or, if the range of pressures for an atmospheric pressure profile is from 984 millibars to 1012 millibars, the vertical axis might range from 980 to 1016. Once the range of values for the vertical axis is known, these values should be evenly spaced along the vertical axis.

Fifth, plot a point directly above each of the tick marks on the horizontal axis of the graph across from the appropriate value on the vertical axis. Then connect these points with a smooth line.

Last, clearly label the profile. Label the axes of the graph clearly. The horizontal axis represents distance across the landscape. Copy the graphic map scale onto the profile to show the horizontal scale. The starting and ending points of the profile should be labeled on the horizontal axis. The vertical axis represents the feature the isolines are showing, such as atmospheric pressure or elevation. On the vertical axis also label the units of measurement associated with that feature, for example, feet or millibars. Give the profile a title. An example is provided in Figure C.1.

FIGURE C.1 PROFILE EXAMPLE

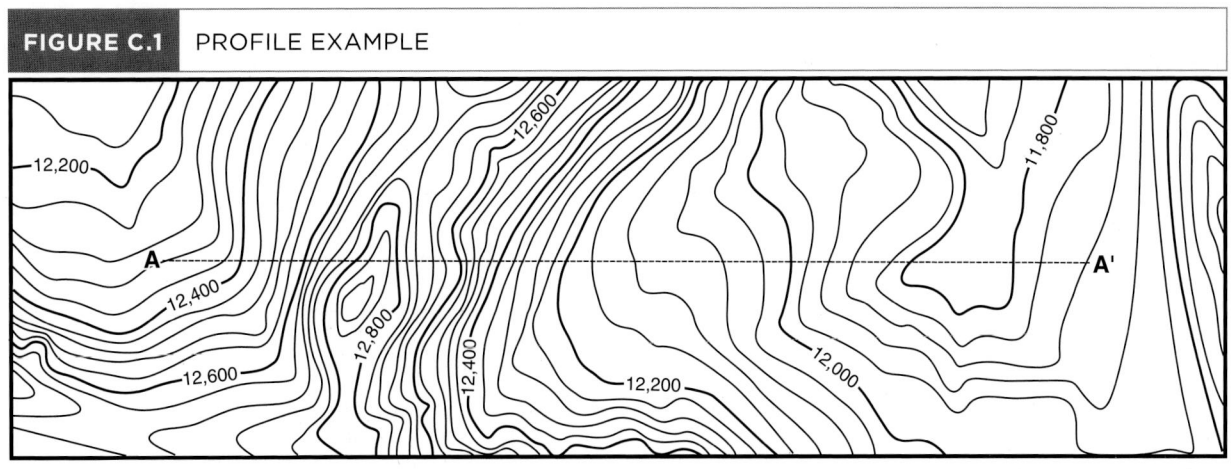

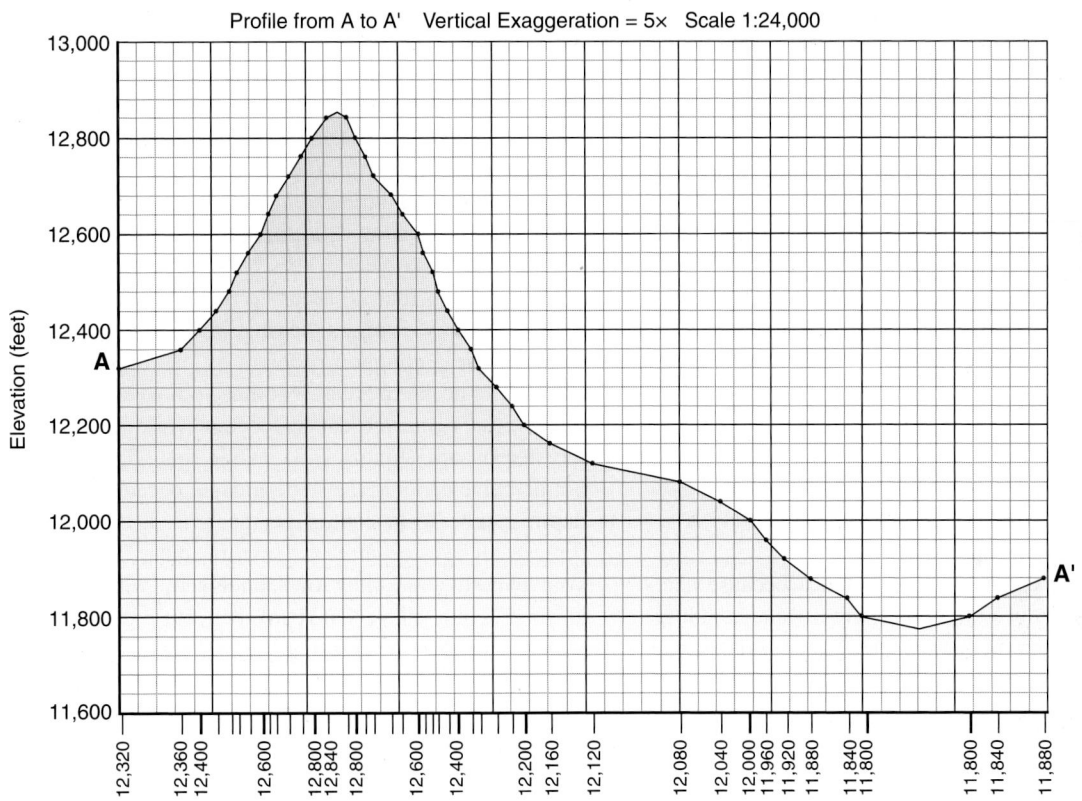

APPENDIX D
USING POCKET STEREOSCOPES

You can align aerial photographs that depict overlapping portions of the ground for three-dimensional viewing, using specially designed lenses mounted in an instrument called a stereoscope. Images viewed through a stereoscope create the illusion of exaggerated depth, which enables you to see topographic details that you might otherwise miss in single, flat photographs. Pocket stereoscopes are lightweight versions of this instrument, and thus are popular as field equipment among various environmental professionals.

A pocket stereoscope consists of a face piece containing the lenses, sometimes mounted on two sets of folding legs. Look at the face piece from above. The large gap is where you place the bridge of your nose while viewing, so that your eyes look properly down through the left and right lenses. Pocket stereoscopes do not work properly if your left eye views through the right lens, or vice versa. You may need to adjust the face piece so that it is appropriate to the distance separating your eyes.

THE PRINCIPLE (AND PROBLEM) IN STEREO VIEWING

Human vision perceives depth and distance because each eye views an object from a slightly different vantage point a few centimeters (the interocular distance) apart. This normal way that you see something distant is called *convergent perspective* (Figure D.1).

For viewing aerial photographs in stereo, we modify the eye-object relationship to simulate the necessary offset of vantage points, insofar as two photographs of the same ground object are taken from vantage points that are separated by many meters along an aircraft flight track. The stereoscope forces your left eye to view the left photograph, and simultaneously your right eye views the right photo (Figure D.2A). We call this unnatural line-of-sight (Figure D.2B) a *central perspective*. Central-perspective viewing feels odd during your initial attempts, because your eye muscles are unused to pulling each eye toward its own separate viewing object.

The reason that it helps to understand the principle behind stereo viewing is that this insight enables beginners to overcome a common problem by use of a simple "card trick." The problem usually is that your eyes try to "cheat" by reverting to convergent perspective while viewing through the stereoscope (Figure D.3). When this happens, you will not perceive depth in the stereo image, because stereo image depth results from central perspective. The solution is to force each eye to "mind its own business." You can force your eyes to view their own photos by vertically placing an obstruction, like a bankcard, beneath the stereoscope to physically block the crossing lines-of-sight (Figure D.4).

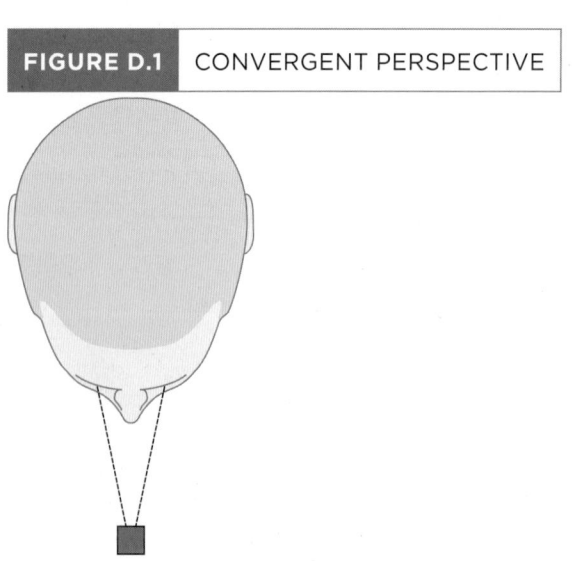

FIGURE D.1 CONVERGENT PERSPECTIVE

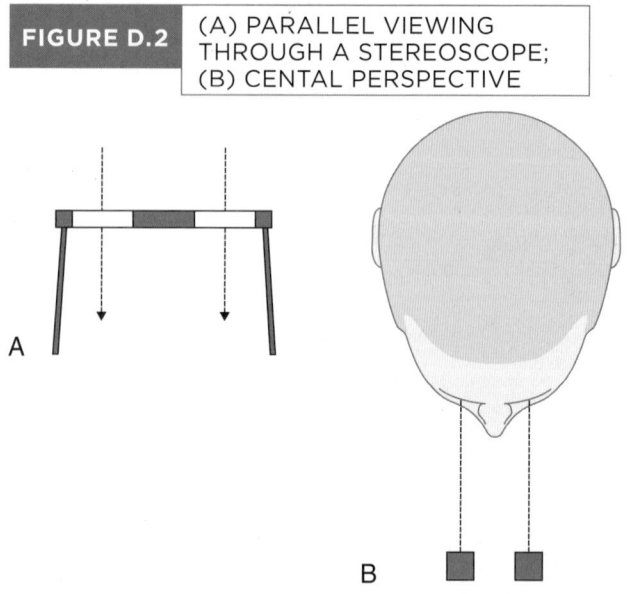

FIGURE D.2 (A) PARALLEL VIEWING THROUGH A STEREOSCOPE; (B) CENTAL PERSPECTIVE

| FIGURE D.3 | ERRONEOUS VIEWING THROUGH A STEREOSCOPE. |

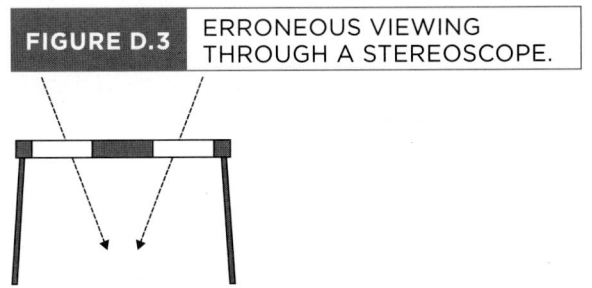

This is the most common cause for initial inability to perceive apparent height and depth.

| FIGURE D.4 | CORRECTING FOR ERRONEOUS PERSPECTIVE |

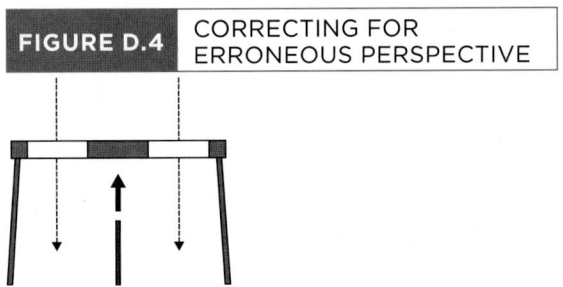

Place a card vertically beneath the nose, so that each eye must view the proper image.

STEREO ACUITY AND THE TEST PATTERN

Researchers design test patterns (Figure D.5) to determine your ability ("acuity") to perceive different apparent depths. This also gives you a chance here to practice setting up and viewing with the stereoscope. What you will see through the scope (Figure D.6) are three nearly identical large circles that contain small numbered objects. Let your eyes relax while looking at the *center image* to see depth/height differences. The objects in the peripheral images (dotted in Figure D.6) will not differ in apparent depth. If you are genuinely blind in one eye, you will not be able to see depth in stereo imagery. However, although there is normal variation in the ability to perceive depth, most persons having two functional eyes can see in stereo to some extent.

Avoid staring through the scope too intently if this is your first try at stereo viewing, because the eyestrain may give you a memorable headache or vertigo. Give your eye muscles a periodic rest. If you have difficulty seeing depth differences in the center image of Figure D.5, try the vertical

| FIGURE D.5 | STEREO ACUITY TEST PATTERN |

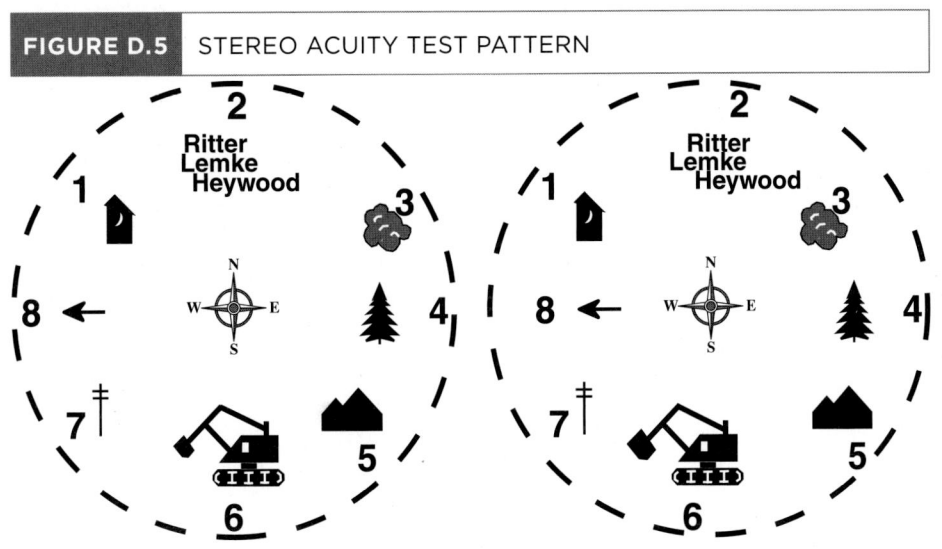

| FIGURE D.6 | STEREOVIEWING EXPECTATION |

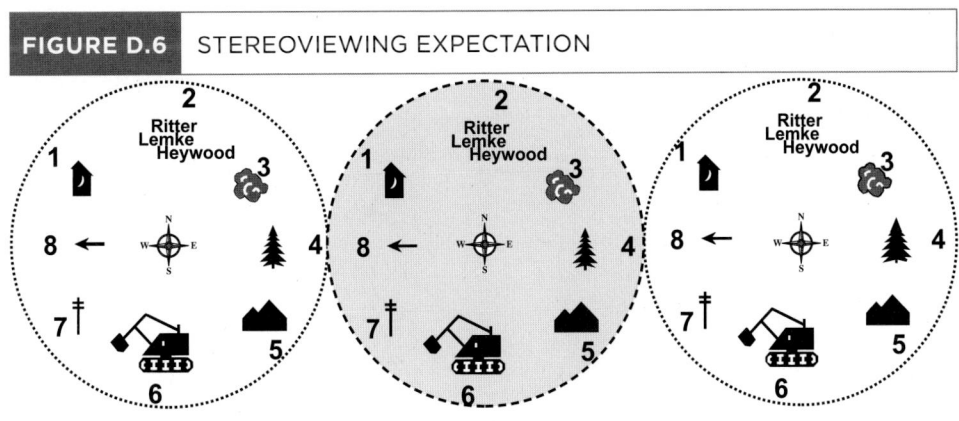

295

card trick described earlier. You should see some of the objects suddenly just pop out at you!

(a) Which of the numbered objects appears highest?
(b) Which of the numbered objects appears deepest?

STEREO AERIAL PHOTOGRAPHY

While it is possible to view separate photographic prints when properly aligned, you will use prealigned photographs because these pose fewer complications for anyone first learning about stereoscopes.

Two photographs comprise most of the air photos in this manual with the boundary between them clearly indicated. Be sure you orient your stereoscope so that (1) the top edge of the face piece parallels the top of the photo, (2) the page remains flat on the tabletop, and (3) the left lens is over the left photo and the right lens is above the right photo. Try to minimize the amount of shadow falling onto the air photos.

Some air photos have two boundaries because they were made out of three photographs. You must view the center photo in these triple images, and one of the other two photos. Unless you have three eyes, you cannot view all three photographs at once!

APPENDIX E
EXERCISE MAPS AND PHOTOS

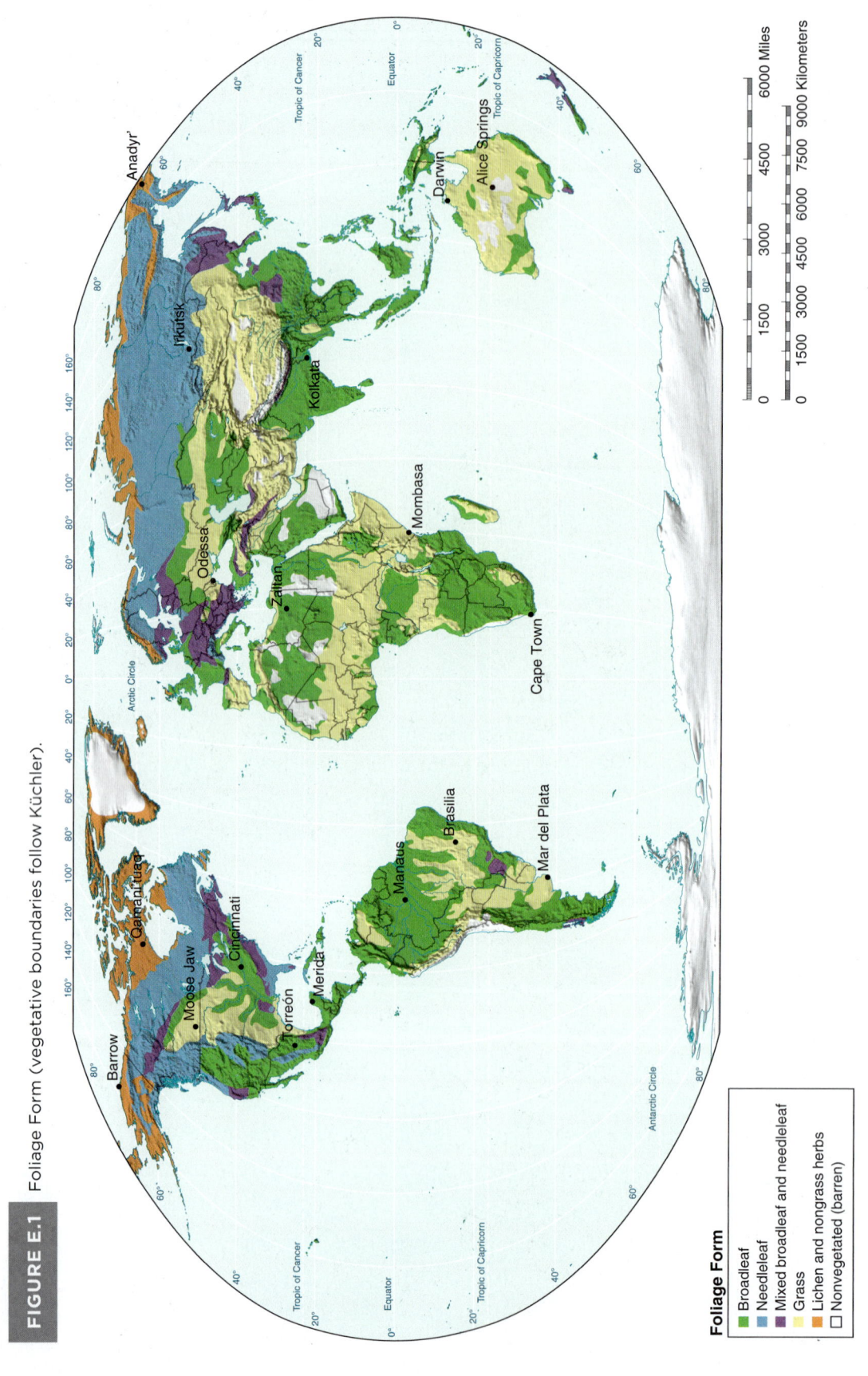

FIGURE E.1 Foliage Form (vegetative boundaries follow Küchler).

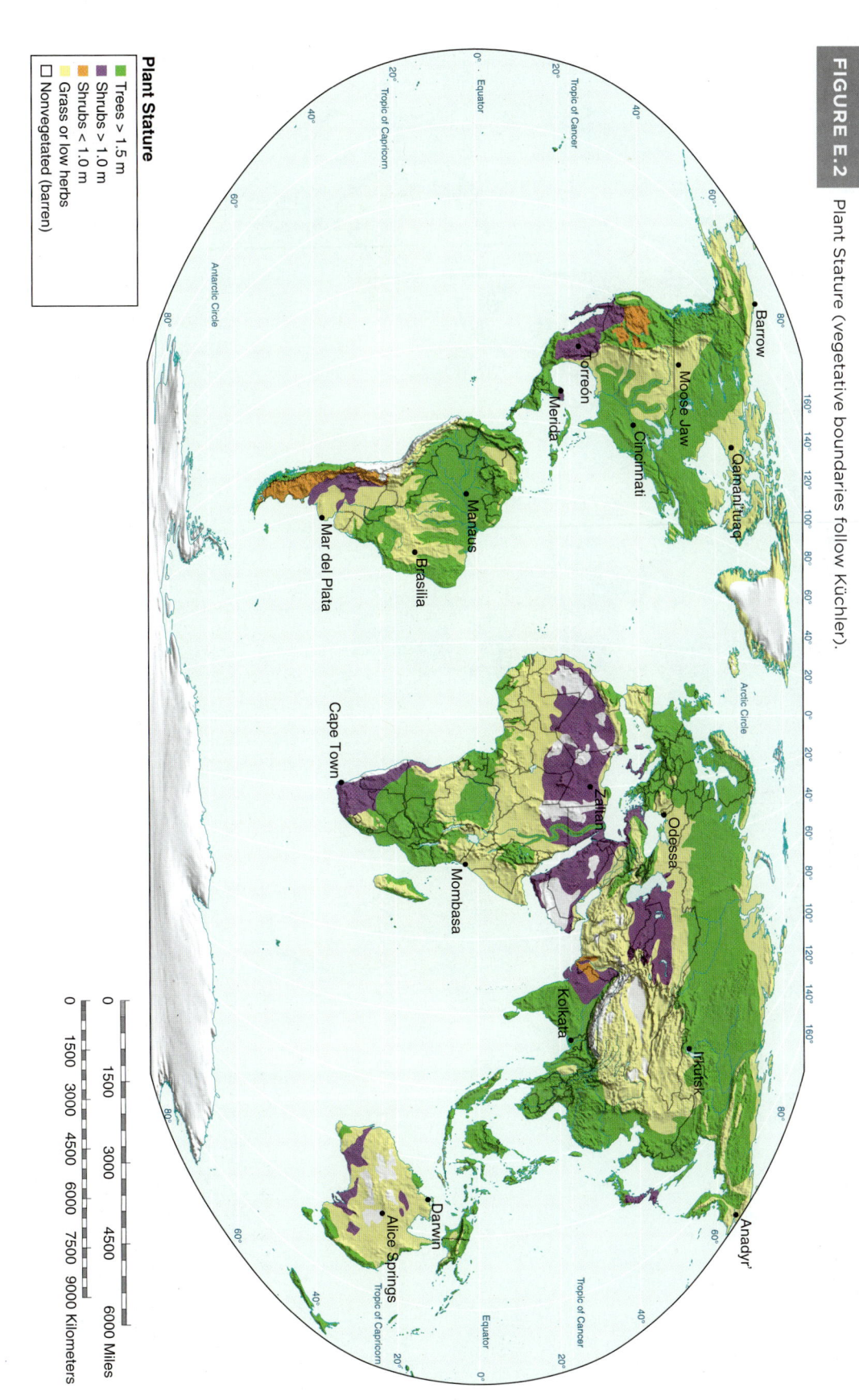

FIGURE E.2 Plant Stature (vegetative boundaries follow Küchler).

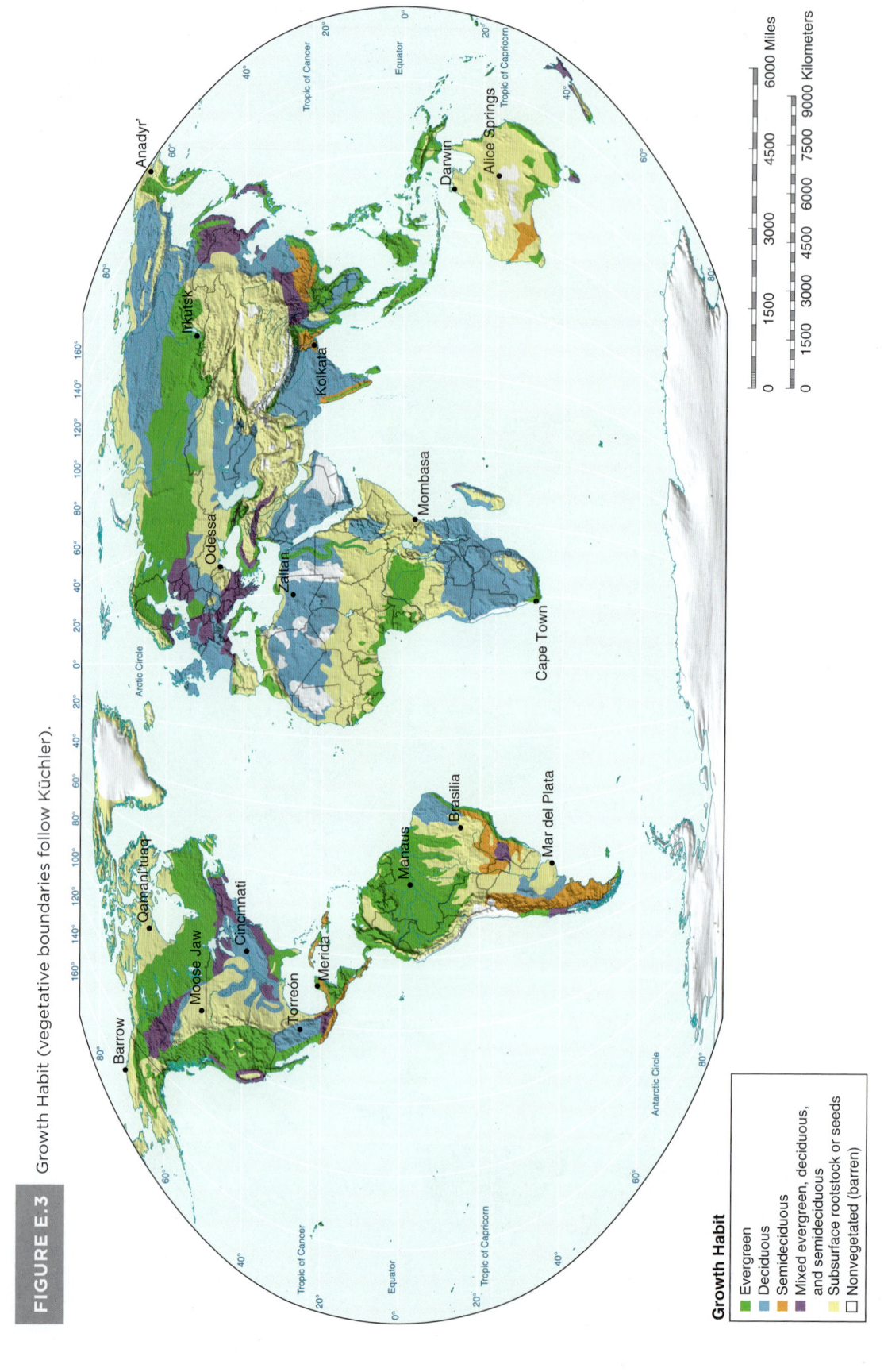

FIGURE E.3 Growth Habit (vegetative boundaries follow Küchler).

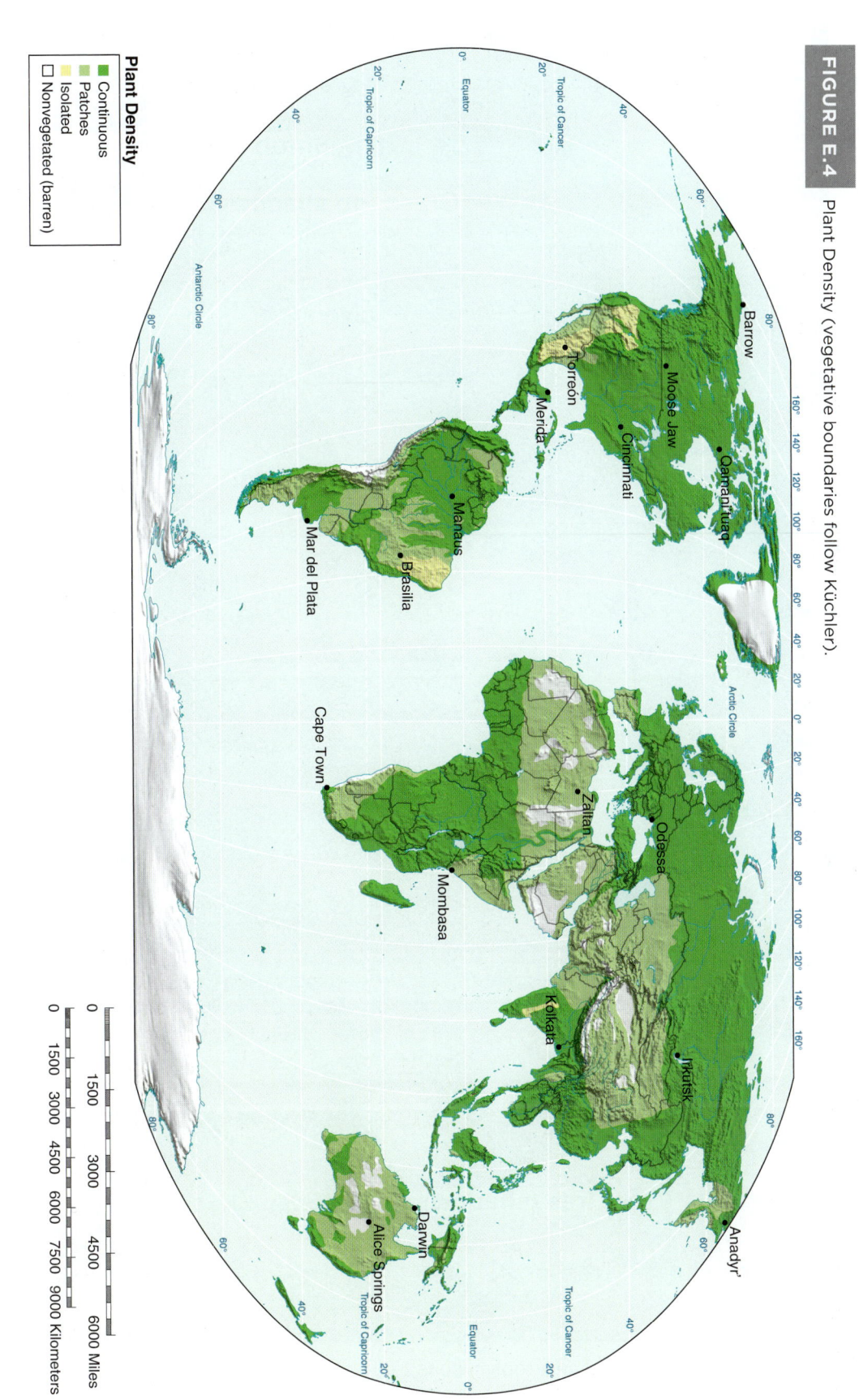

FIGURE E.4 Plant Density (vegetative boundaries follow Küchler).

FIGURE E.5 Aerial Photograph of Devastation Trail Rainforest, 17 March 1987. Local time: 10:44. Scale=1:2,325.

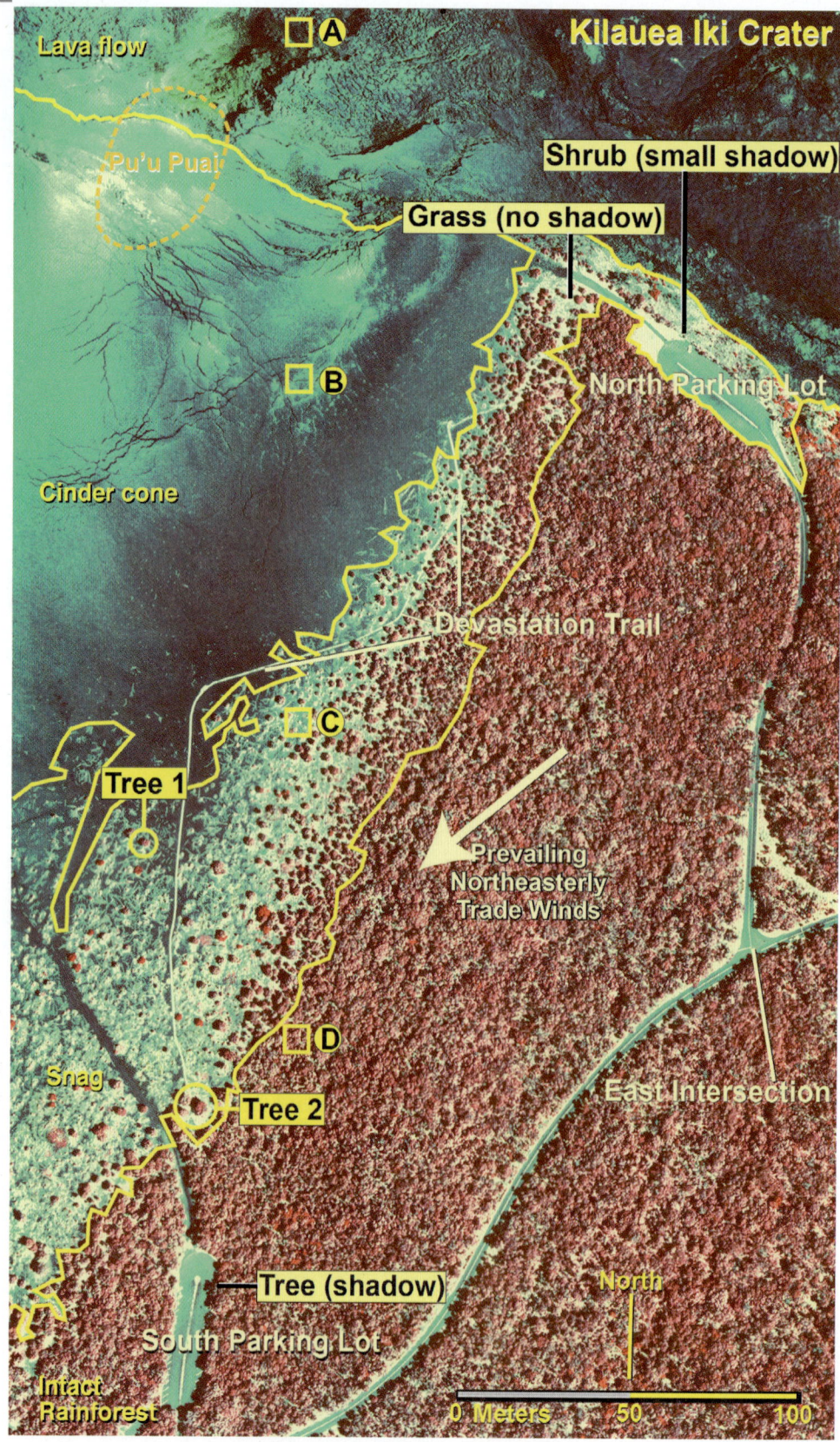

(NASA Ames Research Center AR5870036190046)

FIGURE E.10 USGS Topographic Map Saint Mary, Montana-Alberta 1981. Scale 1:100,000. Contour Interval = 20 meters (from USGS).

FIGURE E.12 USGS Topographic Map Mount St. Helens, Washington 1983. Scale 1:48,000. Contour Interval = 40 feet (from USGS).

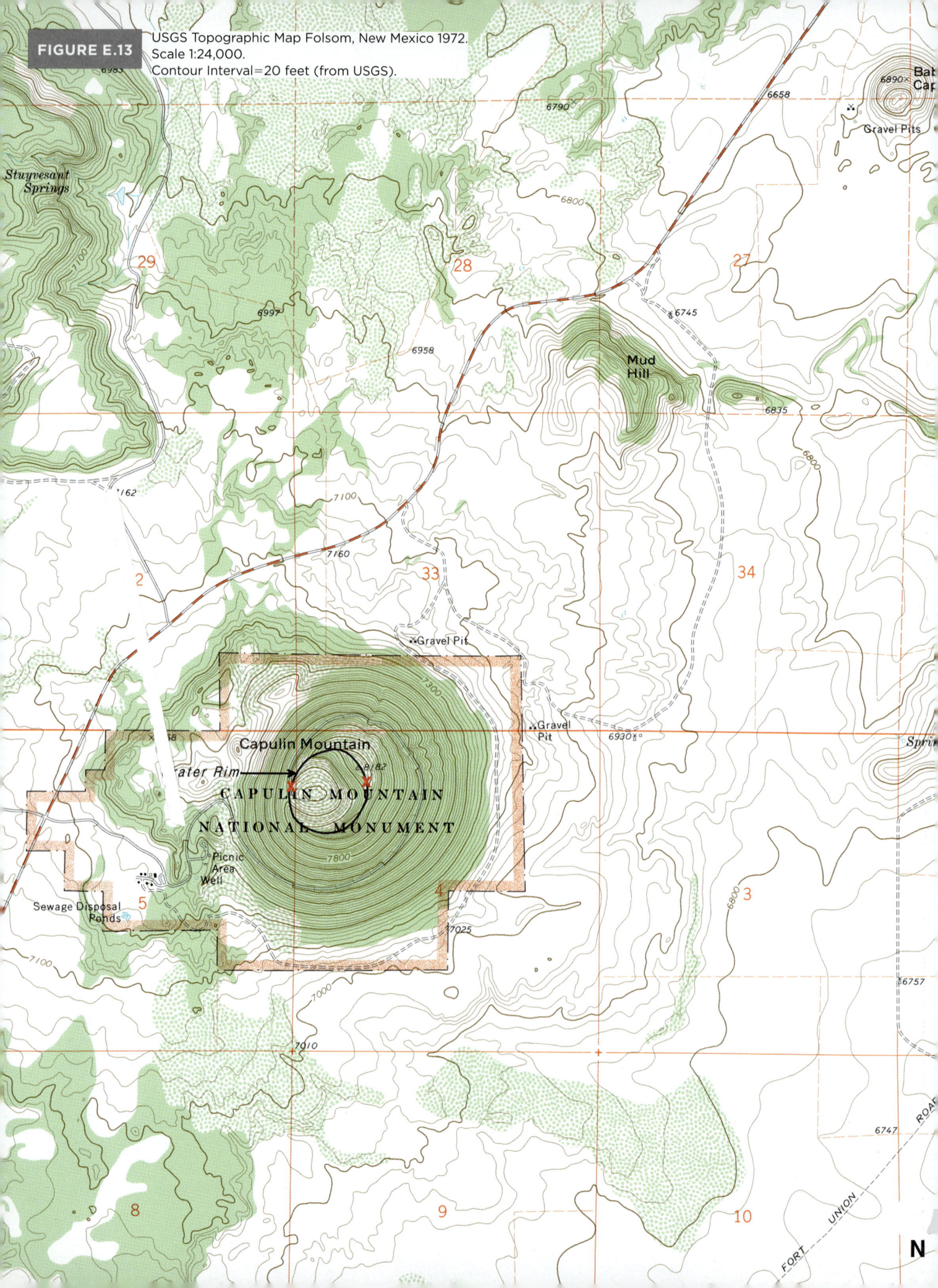

FIGURE E.13 USGS Topographic Map Folsom, New Mexico 1972. Scale 1:24,000. Contour Interval = 20 feet (from USGS).

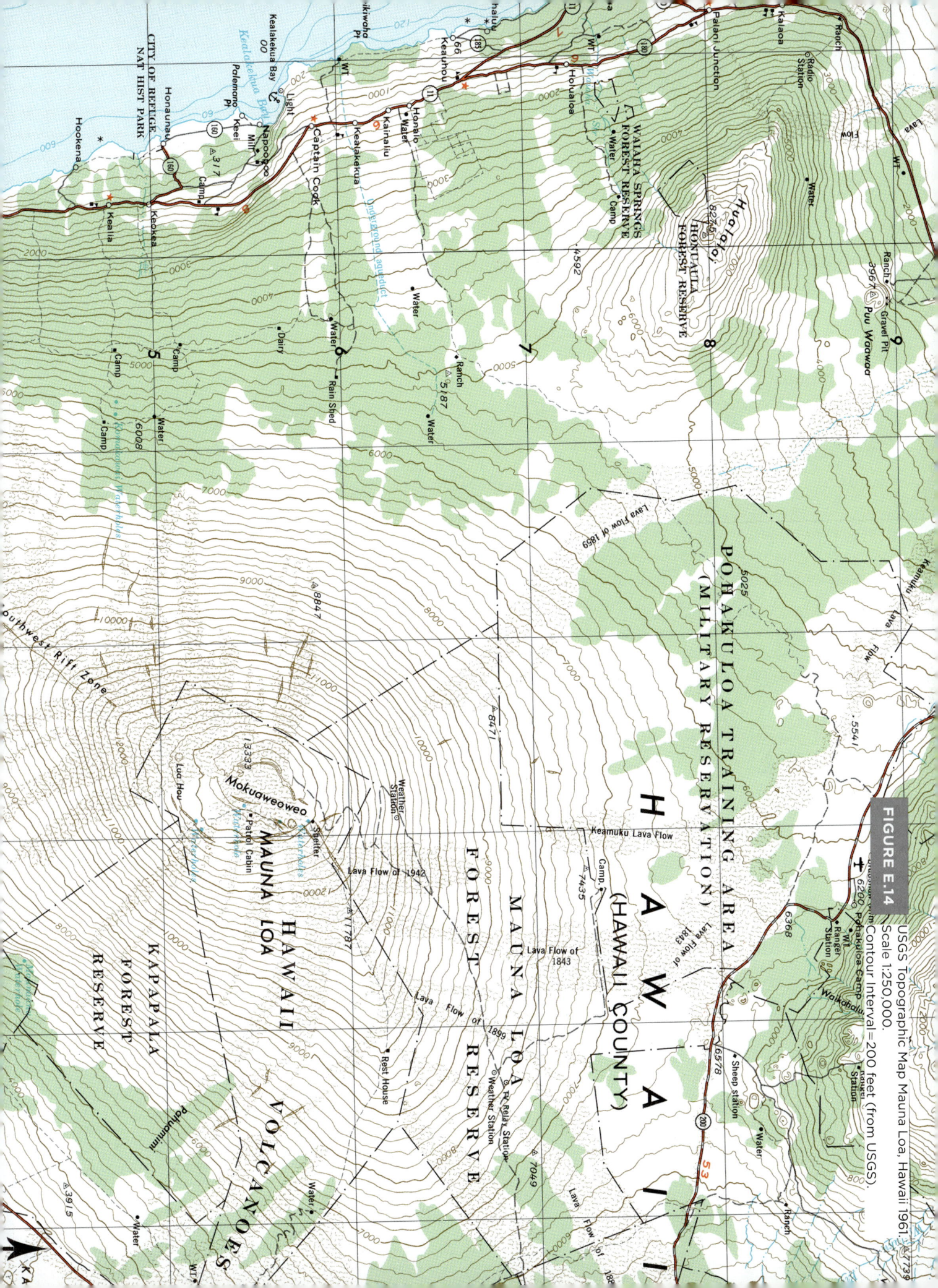

FIGURE E.14 USGS Topographic Map Mauna Loa, Hawaii 1961. Scale 1:250,000. Contour Interval = 200 feet (from USGS).

FIGURE E.15 USGS Topographic Map Ship Rock, New Mexico 1979. Scale 1:24,000. Contour Interval=20 feet (from USGS).

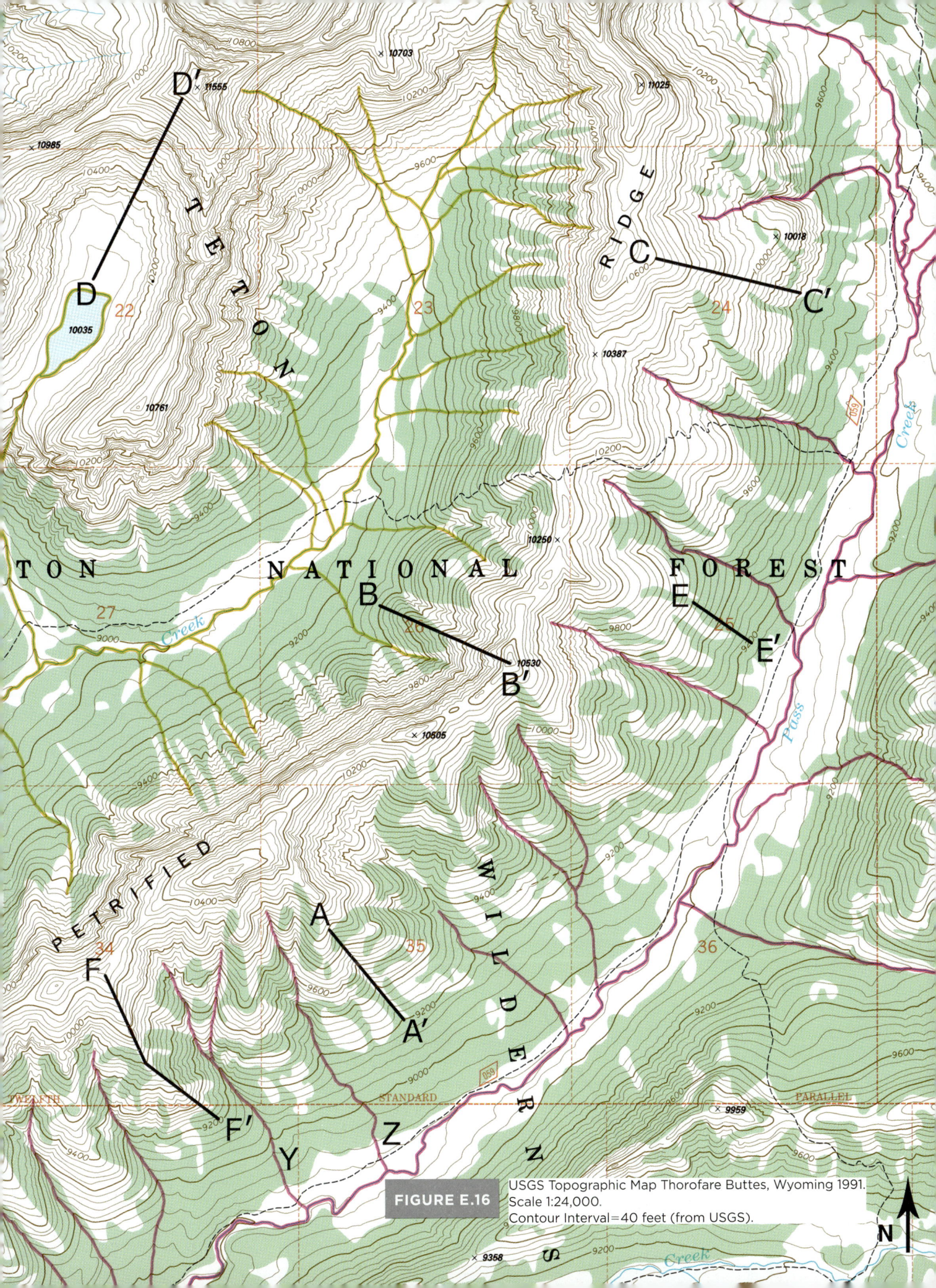

FIGURE E.16 USGS Topographic Map Thorofare Buttes, Wyoming 1991. Scale 1:24,000. Contour Interval=40 feet (from USGS).

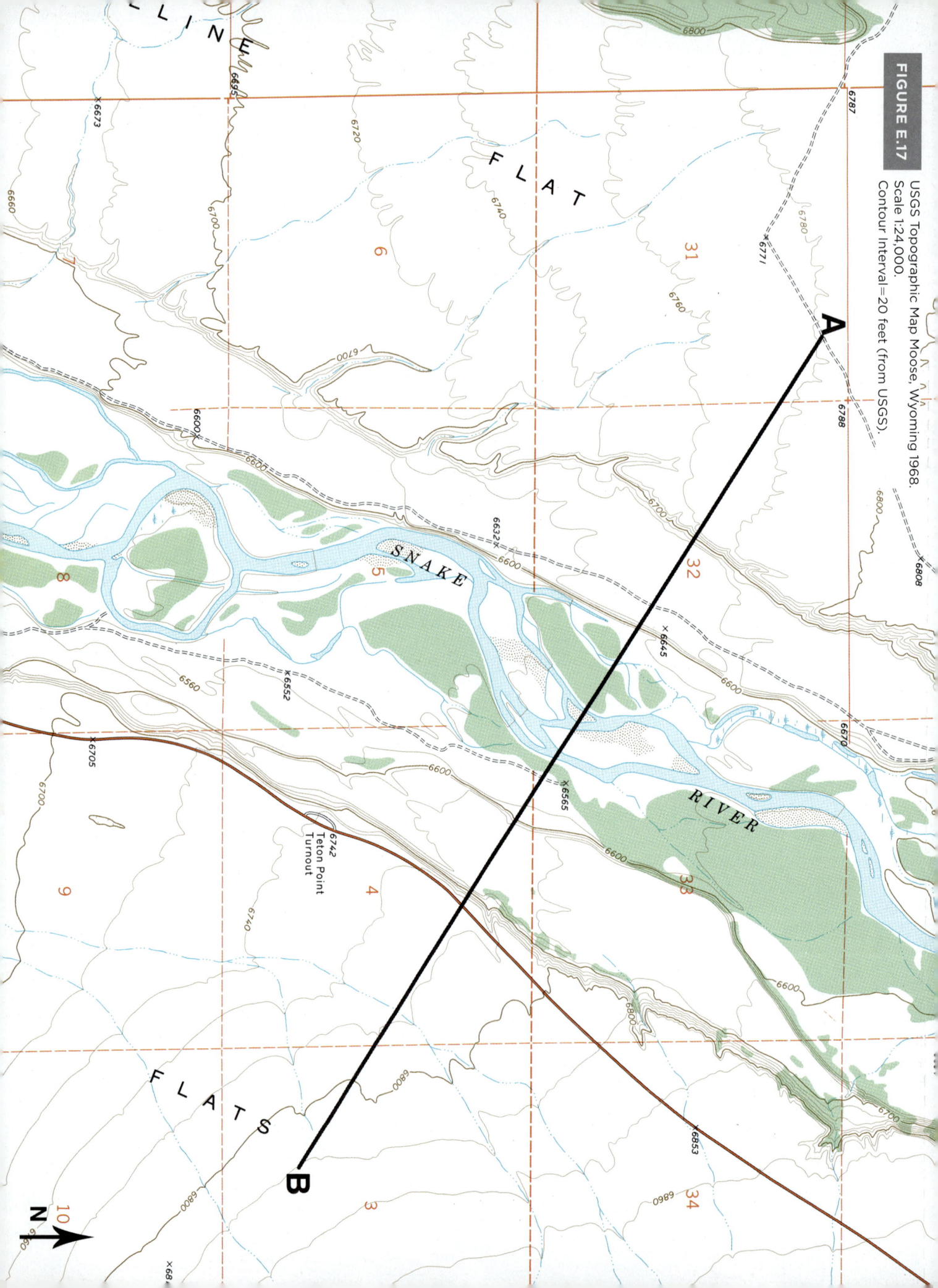

FIGURE E.17 USGS Topographic Map Moose, Wyoming 1968. Scale 1:24,000. Contour Interval=20 feet (from USGS).

EXERCISE 19 † FLUVIAL LANDFORMS

PURPOSE

The purpose of this exercise is to learn about various fluvial landforms and the processes that create them.

LEARNING OBJECTIVES

By the end of this exercise you should be able to

- recognize alluvial terraces and scarps on topographic maps and on topographic profiles;
- identify the edge of a river's floodplain based on changes in topography;
- recognize different stream channel patterns;
- identify landforms associated with meandering rivers; and
- explain some of the ways in which fluvial landscapes change over time.

INTRODUCTION

Rivers are a major force modifying the shape of the landscape through the processes of erosion and deposition. Over time, the dominant process (erosion or deposition) may change, and the magnitude of erosion or deposition may change. Likewise, these processes also change over space (from one place to another).

Base level is the lowest elevation that a stream can attain through erosion and is defined as sea level for rivers flowing into the oceans. Base level for rivers flowing into lakes or other rivers is the elevation of the lake or larger river. All rivers modify the landscape in an attempt to attain a balance between erosion and deposition along the entire length of the river down to base level. When the balance between erosion and deposition is upset, the relative magnitude of these two processes adjusts until a new balance is created. The balance can be upset by a change in base level due either to tectonic activity or a change in sea level. The balance can also be upset by a change in drainage basin conditions and characteristics, which, in turn, affect all the hydrological processes at work within the basin. Adjustments to the magnitude of either erosion and/or deposition create a variety of fluvial landforms. The most obvious landform is the stream channel itself. Other landforms are found within and adjacent to stream channels.

Stream Channel Patterns

The **channel pattern** (channel appearance as viewed from above) for a particular stretch of river provides some indication of the processes at work in that particular place. **Straight channels** are the least common of all channel patterns. Straight channels are most often found in regions where the channel shape is controlled by underlying bedrock and in urban areas where streams have been channelized for the purpose of protecting property along the stream.

Meandering channels are the most common of all channel patterns. Meandering channels are stream channels that swing back and forth across the landscape in a series of S-shaped curves. In regions high above base level, meandering channels may exist in narrow valleys with little or no floodplain, a result of simultaneous vertical and lateral erosion. **Vertical erosion** is erosion down into the landscape, while **lateral erosion** is erosion along one side of the stream channel. In flatter regions closer to base level, meandering channels may result in extensive lateral erosion and deposition, creating wide valleys with well-developed floodplains. Little vertical erosion can occur in rivers close to base level, thus these rivers expend their excess energy in lateral erosion, often resulting in highly meandering rivers.

Braided channels are a third type of channel pattern. Braided channels are river channels that have several distinct ribbons of water flowing in a wide, shallow channel. The individual strands of water are separated by unstable bars and islands. The exact location of each ribbon of water within the channel changes each time storm events or high flows occur. During these higher flow events, many of the existing in-stream bars may get destroyed or reshaped, and new bars will form as flow levels drop. Braided streams are characterized by widespread erosion of the banks, allowing the channel to widen easily, and by steep gradients. Braided streams often occur in environments where discharge fluctuates considerably, and they tend to be dominated by bed-load transport. These conditions allow the formation of ephemeral bars within a wide, shallow main channel, resulting in numerous strands of water that braid.

Floodplain Landforms

The **floodplain** is defined as the flat area of land adjacent to a stream that is periodically inundated as water spills over the channel banks. The edge of a floodplain is marked by an increase in the elevation of the landscape. As rivers flood, water spills onto the floodplain, depositing sediments carried in suspension. These stream-deposited sediments are called **alluvium** and help create the flat area adjacent to the channel known as the floodplain. **Natural levees** may form immediately adjacent to the channel due to deposition of large sediment particles close to the river channel during flood events.

Meandering rivers eroding laterally across the landscape play an important role in the creation of floodplains. Lateral erosion and deposition result from variations in the velocity of streamflow across a stream channel. The highest velocity flow occurs on the outside of meanders and the lowest velocity occurs on the inside of meanders. As a result, lateral erosion is dominant along the outside bank of meanders, causing meanders to move outward toward the edge of the floodplain. This lateral erosion may help widen the floodplain. A steep **cut-bank** may form along the outer bank of meanders when lateral erosion is widening the floodplain. Lateral erosion also moves meanders in a downstream direction. As a result, over time, meanders widen and move downvalley. Deposition is dominant along the inside bank of meanders, forming **point bars.** Within point bar deposits, a series of alternate low ridges (bars) and troughs (swales) are created. This is called **bar and swale topography.** These bars and swales indicate the former channel position and help show how the channel has migrated over time.

Over time, meanders eventually become so curved and extended that the river straightens its course, cutting the meanders off from the main river channel. For a period of time, water flows in the meander as well as in the cutoff channel, forming an **incipient oxbow lake.** Cutoff meanders that are still filled with water but are disconnected from the main channel are called **oxbow lakes.** These lakes eventually fill with sediment, forming **meander scars.** The location of meander scars is evident by the amount of water in the soil and the type of vegetation growing in them (often marsh vegetation). Meander scars can also be seen as semicircular-shaped depressions on the landscape. **Yazoo** streams are small streams flowing in the floodplain to larger rivers. Although these streams would normally join the larger river close to where they flow onto the floodplain, they are prevented from doing so by the natural levees along the larger rivers. The yazoo streams do not have the energy to break through these natural levees. Thus, yazoo streams may flow parallel to the larger river for some distance before they are able to join the larger streams. These landforms associated with floodplains are shown in Figure 19.1.

As the climatic conditions in a region change over time, or as the physiographic and land-use characteristics within a drainage basin change, rivers that at one point in time deposited thick layers of alluvium in their floodplains may stop depositing sediments and start eroding alluvium from the floodplain. For example, during the glacial advances of the last ice age, streams carried high-sediment loads due to the high amounts of sediment contributed by the glaciers. These sediments filled existing river valleys, a process called **aggradation.** Once the glaciers retreated, the sediment load of rivers decreased dramatically, and rivers began to erode down into the sediments deposited during the period of aggradation. This downcutting, or **degradation,** of floodplain alluvium may result in the formation of **alluvial terraces,** or river terraces, and **terrace scarps** (Figure 19.2). Each alluvial terrace marks

FIGURE 19.1 FLOODPLAIN LANDFORMS

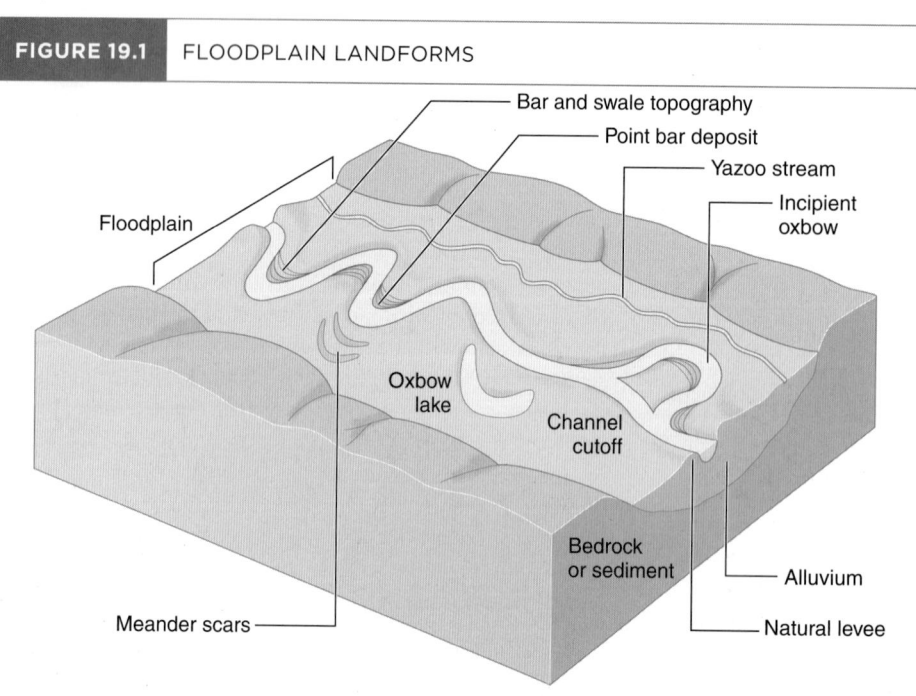

the elevation of a former floodplain, and the terrace scarp (drop-off) marks the edge of a younger floodplain. Rivers may go through alternating periods of aggradation and degradation, depositing alluvium in the floodplain at one point in time and later downcutting and eroding the alluvium.

IMPORTANT TERMS, PHRASES, AND CONCEPTS

base level
channel pattern
straight channels
meandering channels
vertical erosion
lateral erosion
braided channels
floodplain
alluvium
natural levees
meandering rivers
cut-bank
point bar
bar and swale topography
incipient oxbow lake
oxbow lake
meander scar
yazoo stream
aggradation
degradation
alluvial terraces (river terraces)
terrace scarps

FIGURE 19.2 FLUVIAL TERRACES AND TERRACE SCARPS

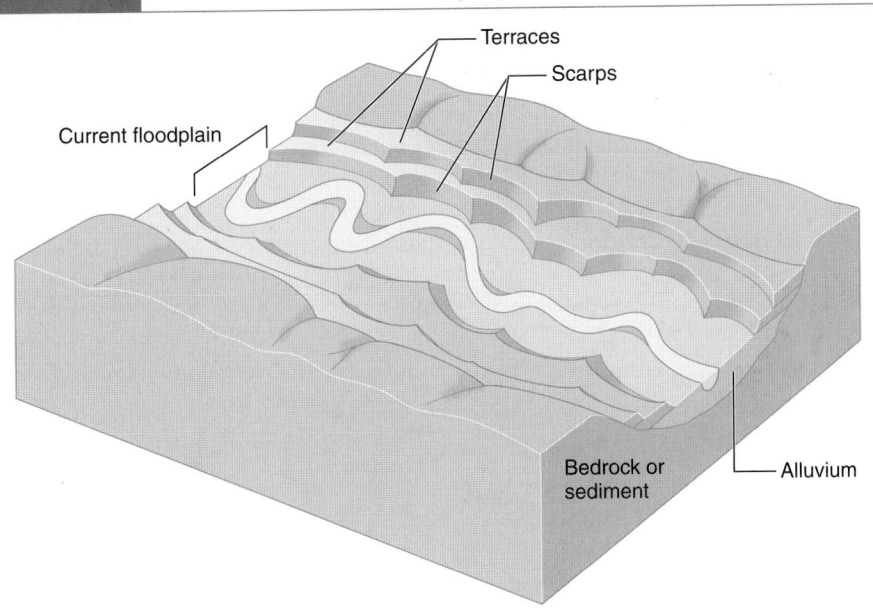

Name: _____ Section: _____

PART 1 ǀ RIVER TERRACES: SNAKE RIVER, WYOMING

Use the topographic map of Moose, Wyoming, (Appendix E, Figure E.17) to answer the following questions.

1. Construct a topographic profile between points A and B on the topographic map of Moose, Wyoming. Graph paper has been provided in Figure 19.3. For this profile to show the necessary detail, all of the contour lines must be used. Using just the index contour lines will not suffice.

2. Label the following features on the topographic profile (Figure 19.3):
   - the channels to the Snake River
   - the width of the current floodplain
   - river (alluvial) terraces
   - terrace scarps

3. Do the terrace elevations on the west side of the river match the terrace elevations on the east side of the river?

4. Based on the number of terraces evident, how many episodes of downcutting occurred in this area?

5. On the topographic map (Appendix E, Figure E.17), mark in red the edges to the current floodplain.

6. Along the profile line, calculate in miles the width of the present floodplain. This may be measured on the map or on the profile; the scale is the same for both. Please show all your work.

7. The water from the Snake River eventually flows into the Pacific Ocean. How high above base level is the Snake River channel where you drew your profile?

8. What type of channel pattern does the Snake River have along the stretch shown here?

249

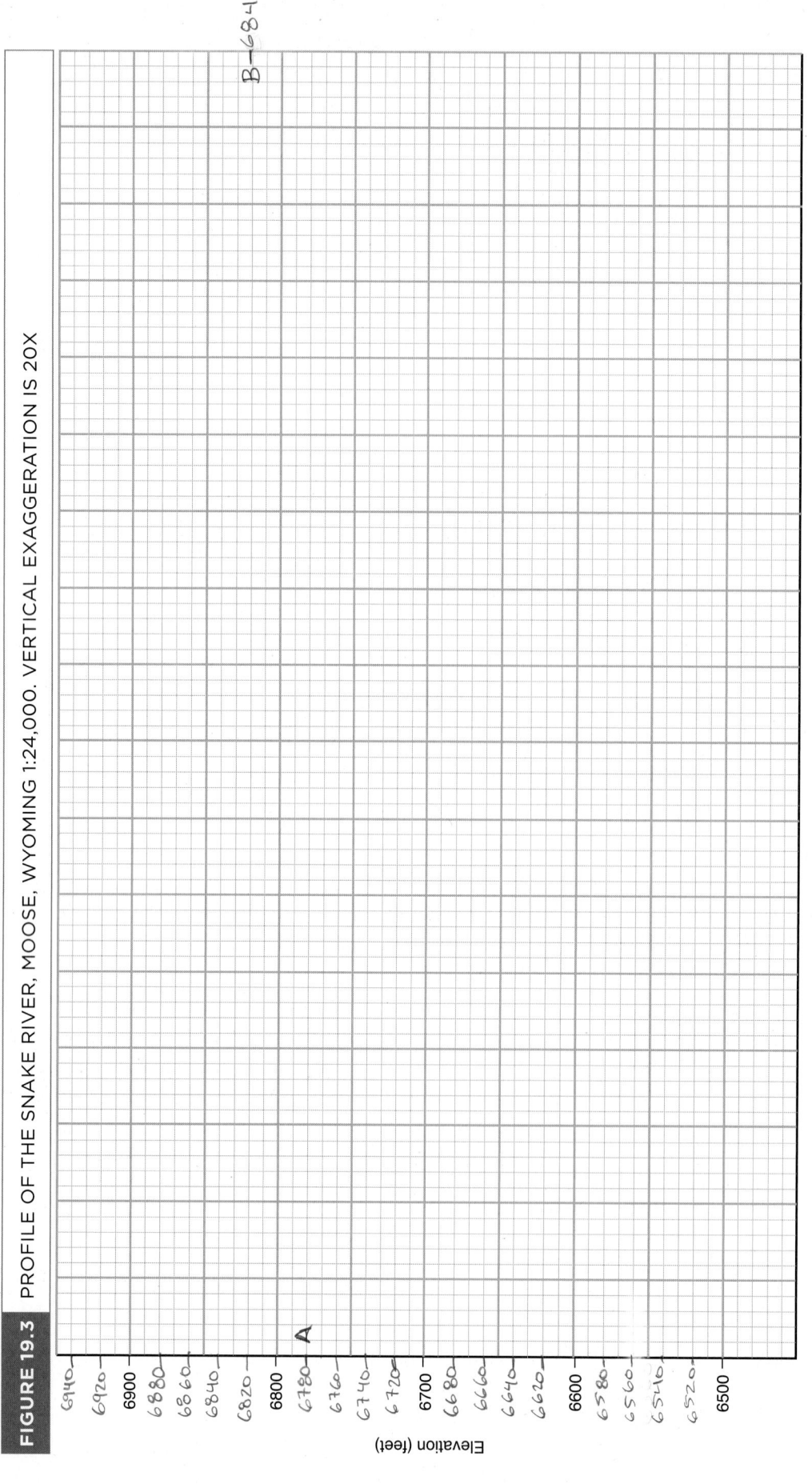

FIGURE 19.3 PROFILE OF THE SNAKE RIVER, MOOSE, WYOMING 1:24,000. VERTICAL EXAGGERATION IS 20X

Name: _____ Section: _____

PART 2 † LANDFORMS OF MEANDERING RIVERS

Use the map of Campti, Louisiana, (Appendix E, Figure E.18) to answer the following questions.

1. On the map of Campti, Louisiana, highlight two or three examples of the following fluvial landforms. In the boxes, create a key for your map, using either different colors or symbols for each feature listed.

   ☐ Point bar deposits or areas of expected deposition along the Red River
   ☐ Areas of expected erosion along the Red River
   ☐ Oxbow lakes
   ☐ Meander scars
   ☐ Yazoo streams
   ☐ Location of the next meander cutoff and formation of a new oxbow lake

2. Question 1 asked you to label the location of the next meander cutoff. Why did you choose this location?

3. a. What is the technical name for the semicircular feature called "Old River"?

   b. Based on your answer for part (a), what type of landform is Smith Island?

   c. Examine the contour lines on Smith Island. These contour lines actually show bar and swale topography; this is the only place where bar and swale topography is revealed by the contour lines. Based on the patterns of these contour lines, approximately how many shifts in the location of the river channel are revealed here?

4. a. Using variations in topography and elevation, estimate the approximate location of the edges (northern and southern) of the floodplain and highlight them on the topographic map. Only portions of the floodplain edge are visible.

   b. A straight line appears on the map just to the west of the town of Campti, running from the north side of the map to the south side of the map and labeled A-B. What is the width of the floodplain along this line, in miles? Please show all your work. Estimate the width of floodplain in Miles.

   Scale:
   1:62,500
   20 feet

   7 Inches on map × 62,500 = 437,500 Inches on Earth
   by
   437,500 ÷ 12 / ft = 36,458 ft. ÷ 5280 ft/miles = 6.9 miles width of floodplain.

5. Water from the Red River eventually flows into the Gulf of Mexico. How high above base level is the Red River?

   90 feet above sea level

6. a. How does the width of the Red River floodplain compare to the width of the Snake River floodplain (from Part 1)?

251

b. Is this difference in floodplain width what you would expect given the difference in elevation above base level? Why or why not?

7. Figure 19.4 is a 1998 air photo of the region shown on the Campti topographic map (Appendix E, Figure E.18). The topographic map was made in 1957.

   a. Locate Old River and Smith Island and highlight them on the air photo. These will serve as reference points to help you relate what you're seeing on the photo to what the topographic map shows.

   b. How has the Red River changed to the north and west of Smith Island in the time interval from when the map was made and the photo taken?

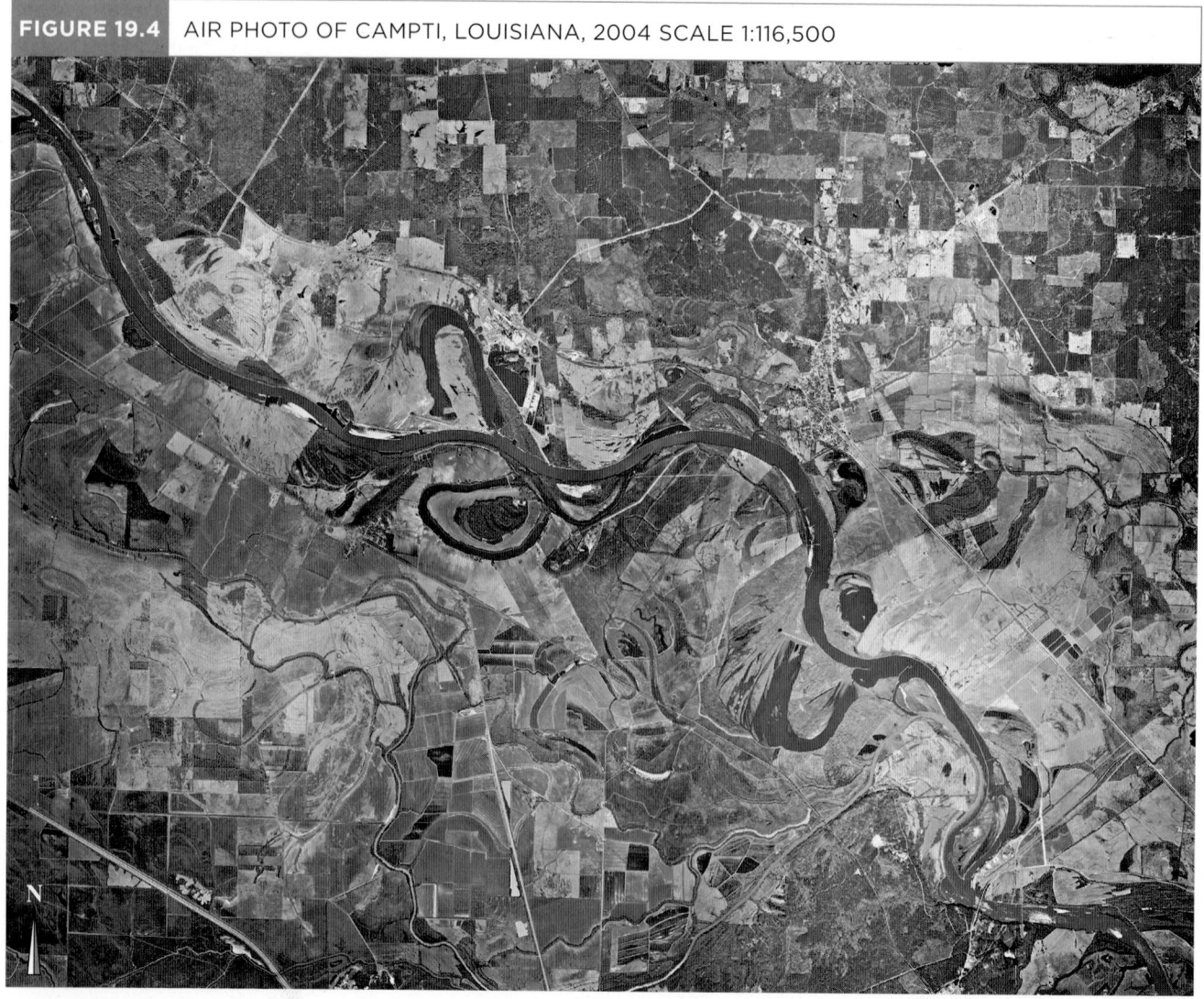

FIGURE 19.4 AIR PHOTO OF CAMPTI, LOUISIANA, 2004 SCALE 1:116,500

Source: Maps courtesy of the USGS.

c. Locate the city of Campti on the air photo. How has the river changed to the south of Campti?

d. Did a cutoff occur at the location you marked for question 1?

e. Locate two new oxbow lakes or incipient oxbow lakes on the air photo and highlight them.

8. Compare the length of the Red River on the topographic map (Appendix E, Figure E.18) to its new length on the air photo (Figure 19.4).

   a. Use a piece of string or a map measurer to measure the length of the river in miles. Note that the scale for the air photo is not equal to the scale for the topographic map.

   River length on topographic map (miles): _____9.6 miles_____
   1957 map = 9.75 Inches
   9.75 In × 62.500 = 9.6 miles

   River length on air photo (miles): _____6.4 miles_____
   3.5 Inches

   b. Assuming the elevation of the river at its northernmost point on the map and air photo has not changed over time, and assuming the elevation of the river at its southernmost point on the map and air photo has not changed over time, how will this change in length affect the river gradient?

   c. What impact should this change in gradient have on flow velocity and on the river's ability to transport sediment? Should there be more erosion or more deposition in 2004 than at the time the map was made? Explain.

   If gradient ↑ the flow rate of water increases also is able to transport more sediment.

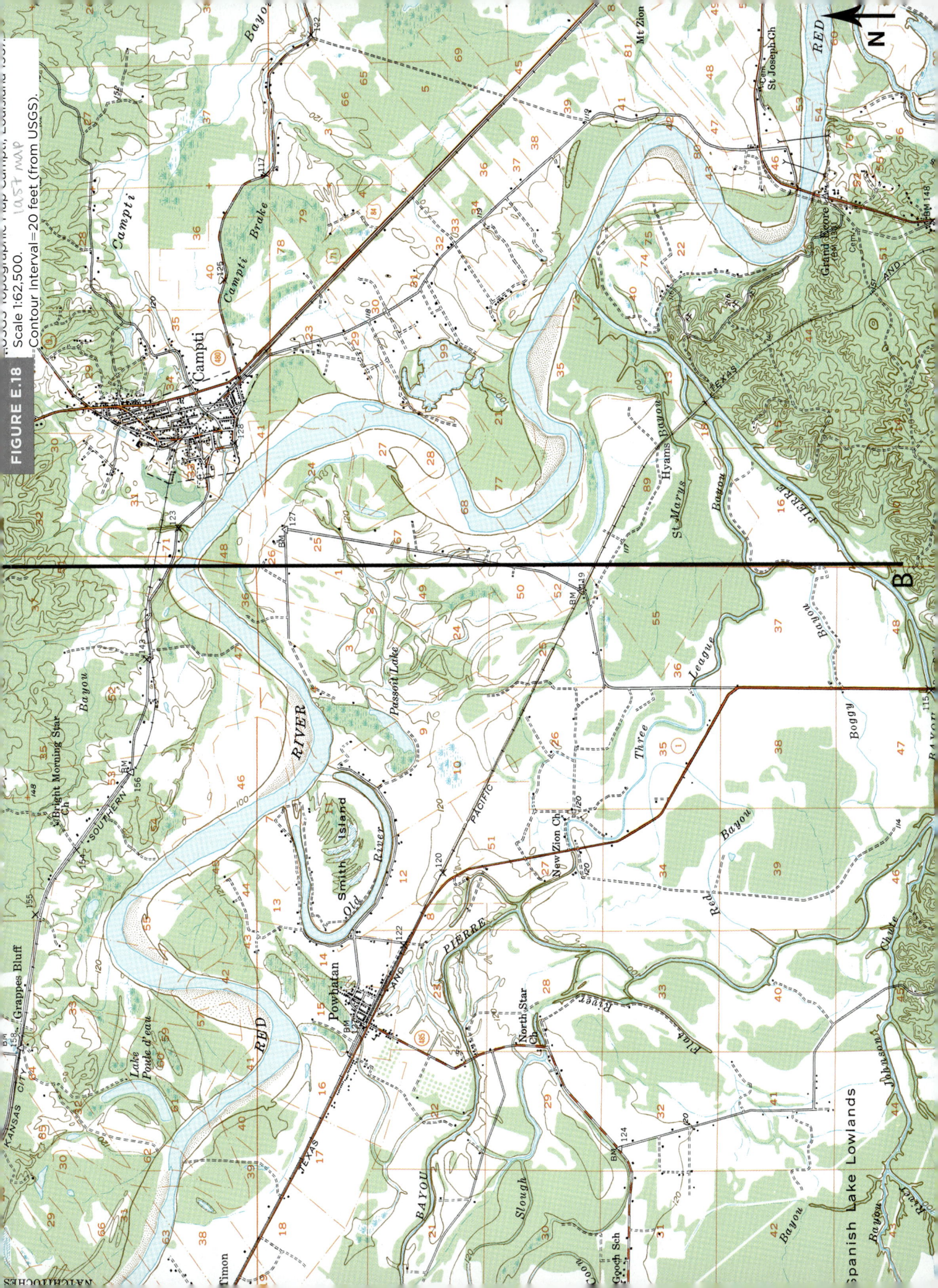

FIGURE E.18 USGS Topographic Map Campti, Louisiana 1957. Scale 1:62,500. Contour Interval = 20 feet (from USGS).

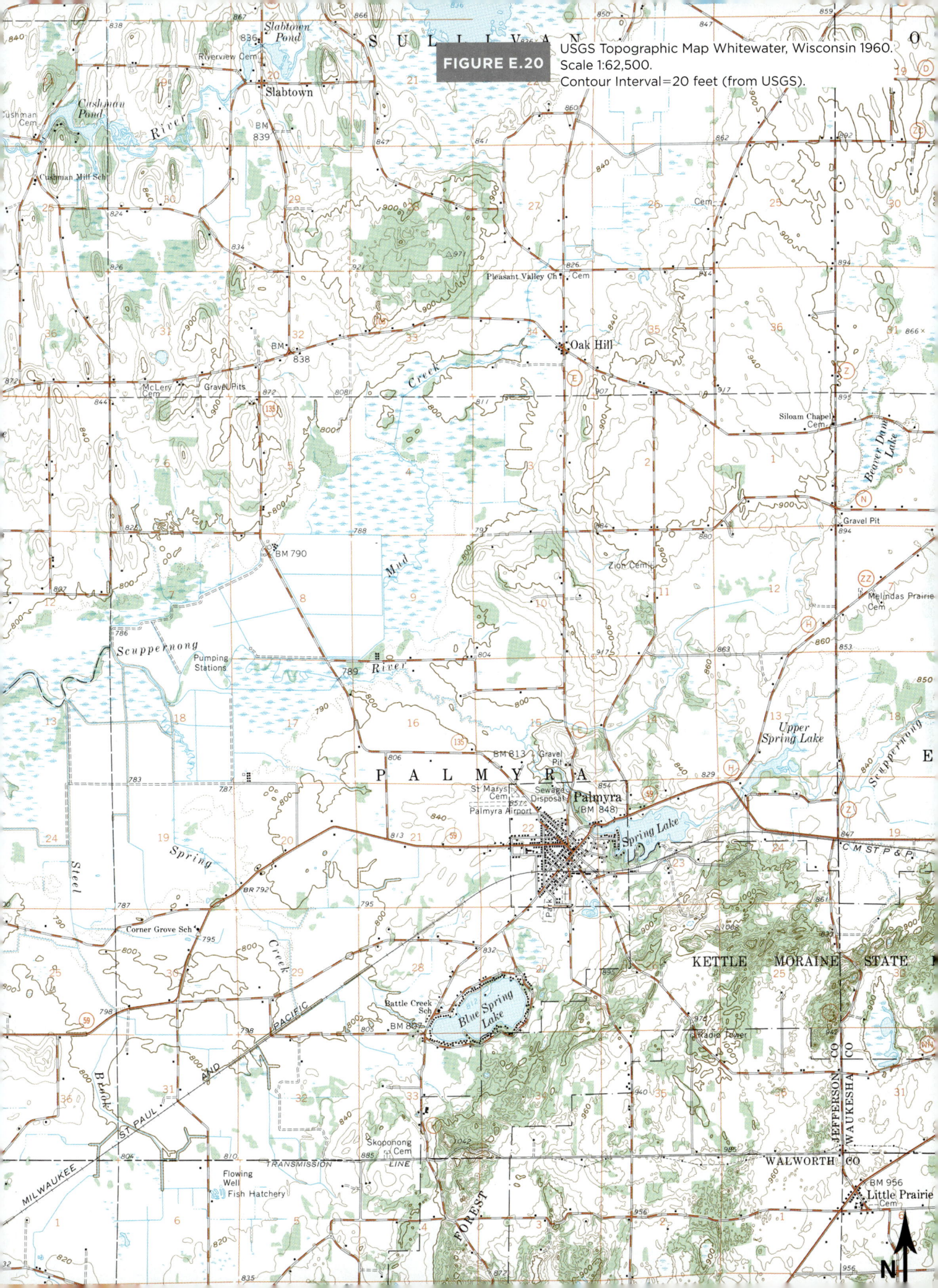
FIGURE E.20 USGS Topographic Map Whitewater, Wisconsin 1960. Scale 1:62,500. Contour Interval=20 feet (from USGS).

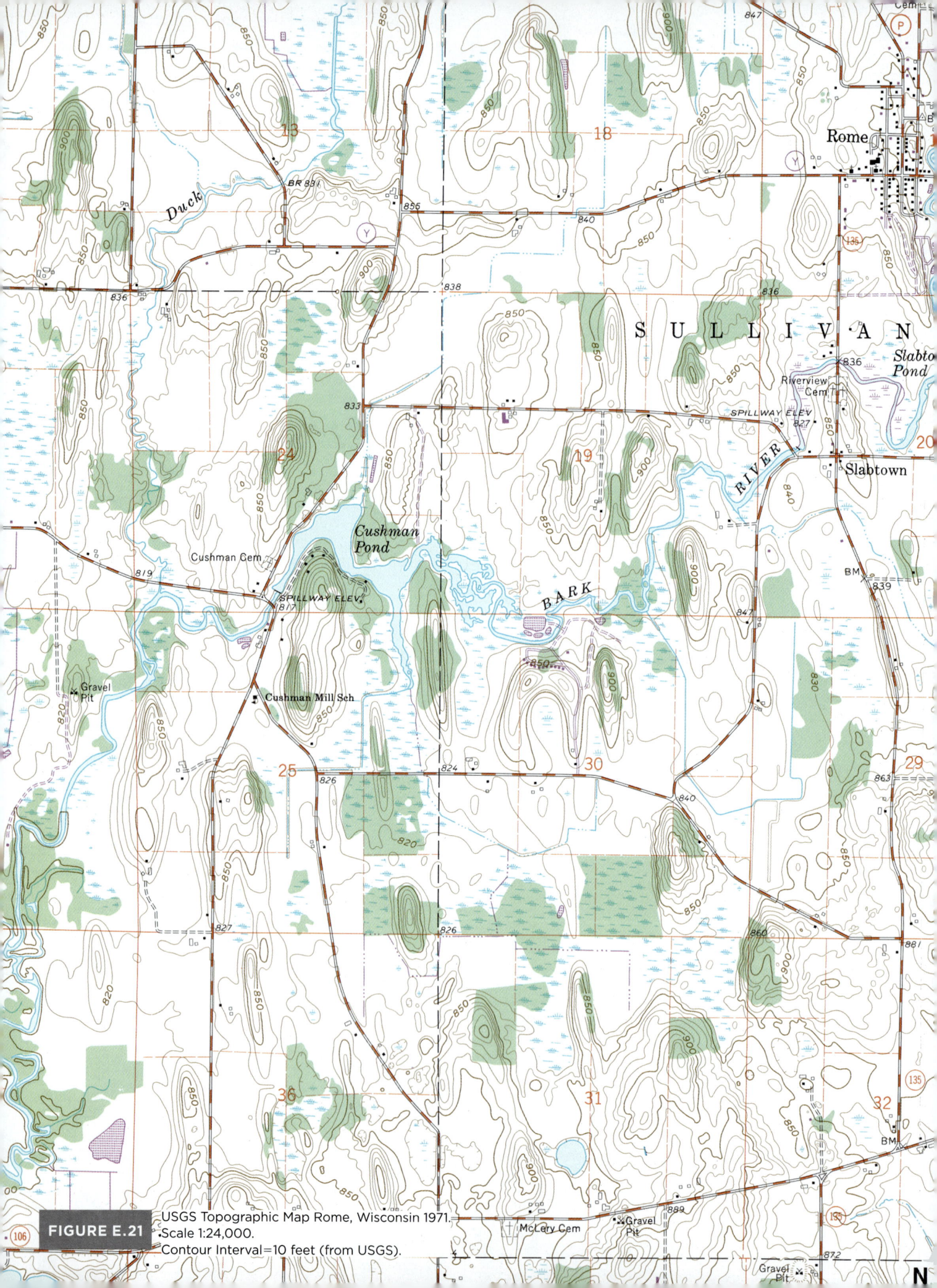

FIGURE E.21 USGS Topographic Map Rome, Wisconsin 1971. Scale 1:24,000. Contour Interval = 10 feet (from USGS).

FIGURE E.22 USGS Topographic Map Portions of Provincetown and North Truro 1972 (original scale 1:25,000). Scale 1:37,879. Contour Interval = 10 feet (from USGS).

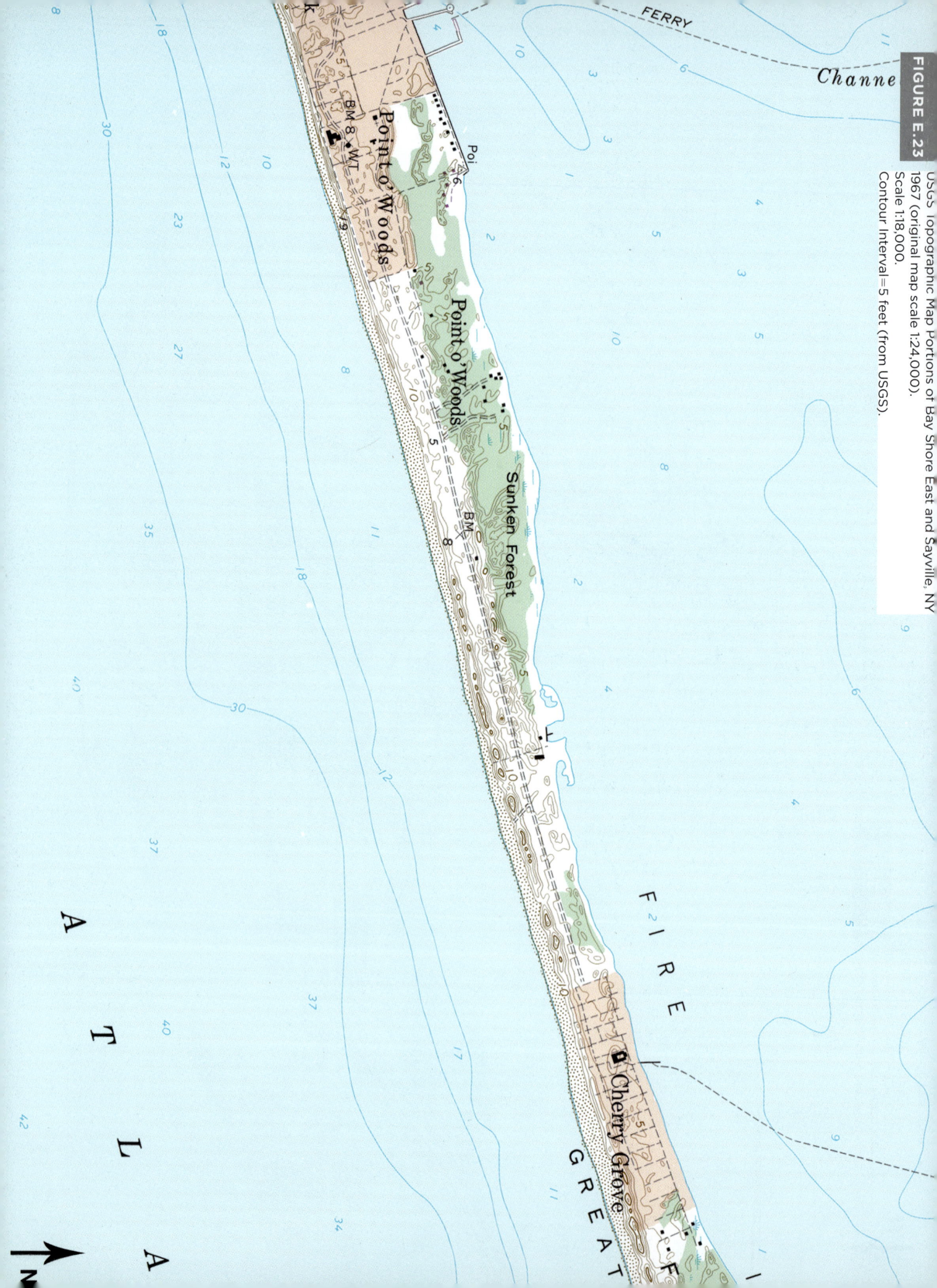

FIGURE E.23 USGS Topographic Map Portions of Bay Shore East and Sayville, NY 1967 (original map scale 1:24,000). Scale 1:18,000. Contour Interval=5 feet (from USGS).

FIGURE E.24 USGS Topographic Map Redondo Beach, California 1972. Scale 1:24,000. Contour Interval=20 feet (from USGS).

FIGURE E.25 — Key to Topographic Map Symbols

CONTROL DATA AND MONUMENTS

Aerial photograph roll and frame number*   3-20

Horizontal control

Third order or better, permanent mark	Neace △ Neace ⊕
With third order or better elevation	BM △ 45.1 Pike ⊕ BM 45.1
Checked spot elevation	△ 19.5
Coincident with section corner	Cactus △ Cactus ⊕
Unmonumented*	+

Vertical control

Third order or better, with tablet	BM × 16.3
Third order or better, recoverable mark	× 120.0
Bench mark at found section corner	BM + 18.6
Spot elevation	× 5.3

Boundary monument

With tablet	BM □ 21.6 BM ⊕ 71
Without tablet	□ 171.3
With number and elevation	67 □ 301.1

U.S. mineral or location monument   ▲

CONTOURS

Topographic
- Intermediate
- Index
- Supplementary
- Depression
- Cut; fill

Bathymetric
- Intermediate
- Index
- Primary
- Index Primary
- Supplementary

BOUNDARIES
- National
- State or territorial
- County or equivalent
- Civil township or equivalent
- Incorporated city or equivalent
- Park, reservation, or monument
- Small park

LAND SURVEY SYSTEMS

U.S. Public Land Survey System
- Township or range line
- Location doubtful
- Section line
- Location doubtful
- Found section corner; found closing corner
- Witness corner; meander corner   WC / MC

Other land surveys
- Township or range line
- Section line
- Land grant or mining claim; monument
- Fence line

SURFACE FEATURES
- Levee
- Sand or mud area, dunes, or shifting sand
- Intricate surface area
- Gravel beach or glacial moraine
- Tailings pond

MINES AND CAVES
- Quarry or open pit mine   ✕
- Gravel, sand, clay, or borrow pit   ✕
- Mine tunnel or cave entrance
- Prospect; mine shaft   ×   ▪
- Mine dump
- Tailings

VEGETATION
- Woods
- Scrub
- Orchard
- Vineyard
- Mangrove

GLACIERS AND PERMANENT SNOWFIELDS
- Contours and limits
- Form lines

MARINE SHORELINE

Topographic maps
- Approximate mean high water
- Indefinite or unsurveyed

Topographic-bathymetric maps
- Mean high water
- Apparent (edge of vegetation)

---

*Provisional Edition maps only

Provisional Edition maps were established to expedite completion of the remaining large scale topographic quadrangles of the conterminous United States. They contain essentially the same level of information as the standard series maps. This series can be easily recognized by the title "Provisional Edition" in the lower right hand corner.

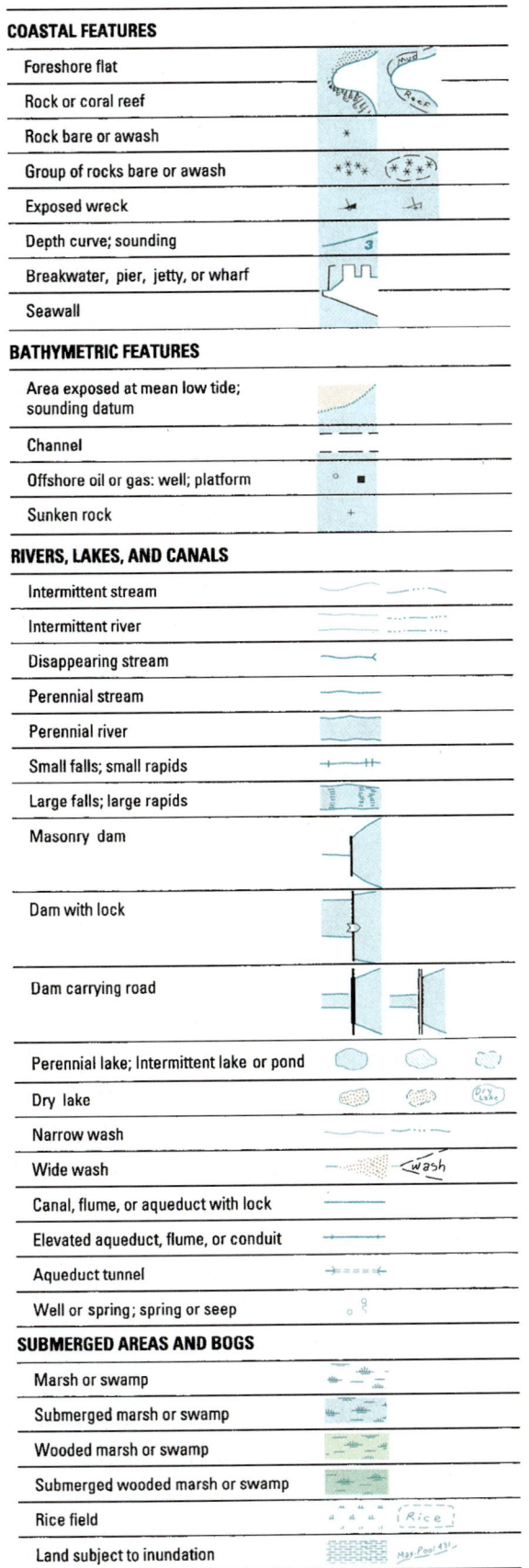

COASTAL FEATURES
Foreshore flat	
Rock or coral reef	
Rock bare or awash	
Group of rocks bare or awash	
Exposed wreck	
Depth curve; sounding	
Breakwater, pier, jetty, or wharf	
Seawall	

BATHYMETRIC FEATURES
Area exposed at mean low tide; sounding datum	
Channel	
Offshore oil or gas: well; platform	
Sunken rock	

RIVERS, LAKES, AND CANALS
Intermittent stream	
Intermittent river	
Disappearing stream	
Perennial stream	
Perennial river	
Small falls; small rapids	
Large falls; large rapids	
Masonry dam	
Dam with lock	
Dam carrying road	
Perennial lake; Intermittent lake or pond	
Dry lake	
Narrow wash	
Wide wash	
Canal, flume, or aqueduct with lock	
Elevated aqueduct, flume, or conduit	
Aqueduct tunnel	
Well or spring; spring or seep	

SUBMERGED AREAS AND BOGS
Marsh or swamp	
Submerged marsh or swamp	
Wooded marsh or swamp	
Submerged wooded marsh or swamp	
Rice field	
Land subject to inundation	

BUILDINGS AND RELATED FEATURES
Building	
School; church	
Built-up Area	
Racetrack	
Airport	
Landing strip	
Well (other than water); windmill	
Tanks	
Covered reservoir	
Gaging station	
Landmark object (feature as labeled)	
Campground; picnic area	
Cemetery: small; large	

ROADS AND RELATED FEATURES

Roads on Provisional edition maps are not classified as primary, secondary, or light duty. They are all symbolized as light duty roads.

Primary highway	
Secondary highway	
Light duty road	
Unimproved road	
Trail	
Dual highway	
Dual highway with median strip	
Road under construction	
Underpass; overpass	
Bridge	
Drawbridge	
Tunnel	

RAILROADS AND RELATED FEATURES
Standard gauge single track; station	
Standard gauge multiple track	
Abandoned	
Under construction	
Narrow gauge single track	
Narrow gauge multiple track	
Railroad in street	
Juxtaposition	
Roundhouse and turntable	

TRANSMISSION LINES AND PIPELINES
Power transmission line: pole; tower	
Telephone line	
Aboveground oil or gas pipeline	
Underground oil or gas pipeline	

APPENDIX F
THEMATIC WORLD MAPS

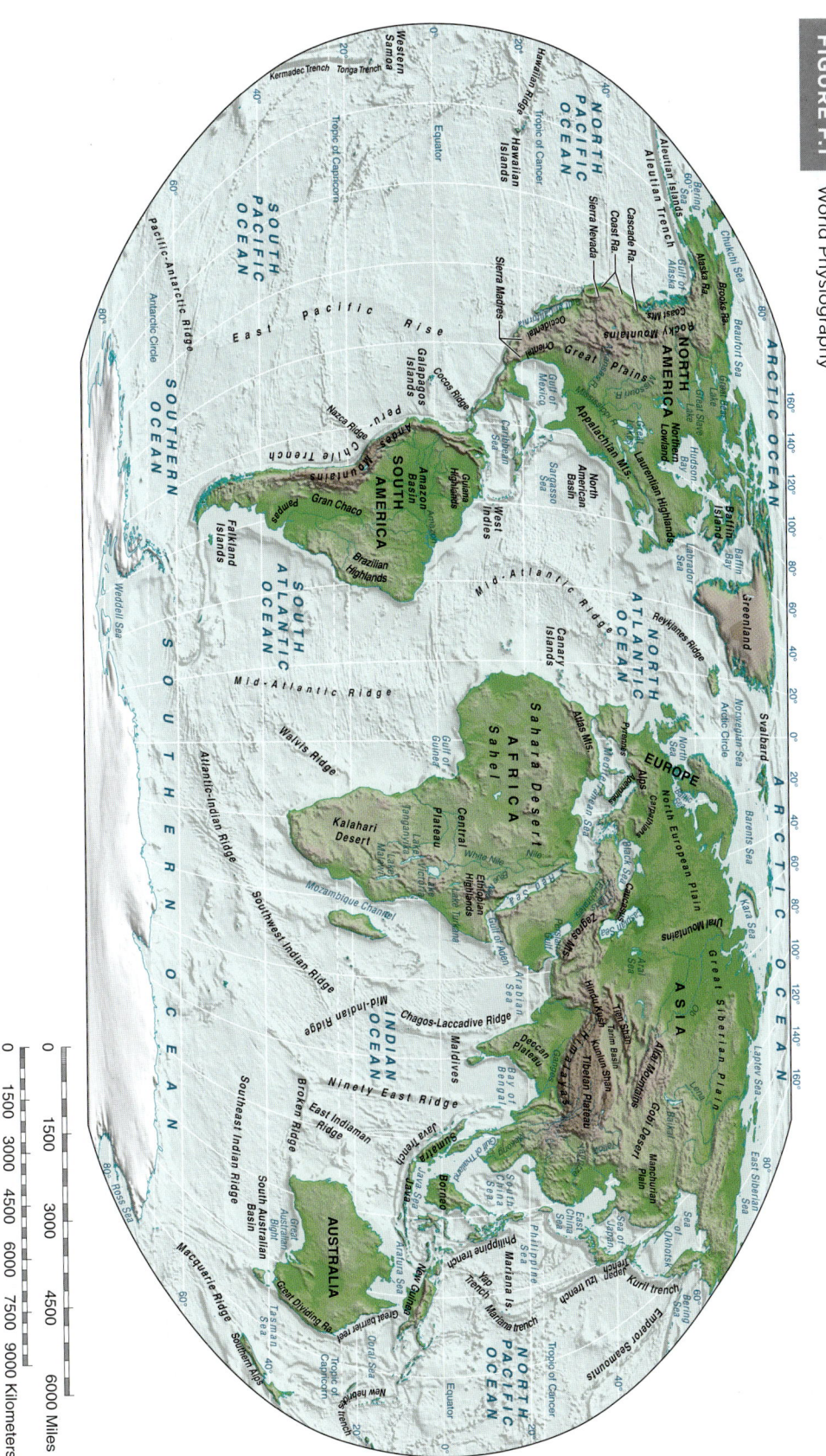

FIGURE F.1 World Physiography

Source: Adapted from Allen 2003

FIGURE F.2 World Climatic Regions

0 1500 3000 4500 6000 Miles

0 1500 3000 4500 6000 7500 9000 Kilometers

Climatic Regions

Tropical moist climates
- Rain forest (Af)
- Monsoon (Am)
- Savanna (Aw)

Dry climates
- Warm desert (BWh)
- Warm steppe (BSh)
- Cool desert (BWk)
- Cool steppe (BSk)

Midlatitude climates
- Summer dry or Mediterranean (Csa, Csb)
- Humid subtropical (Cfa, Cwa, Cwb)
- Marine west coast (Cfb, Cfc)
- Humid continental warm summer (Dfa, Dwa)
- Humid continental cool summer (Dfb, Dwb)
- Subarctic (Dfc, Dfd, Dwc, Dwd)

Polar climates
- Tundra (ET)
- Ice cap (EF)

Azonal climates
- Undifferentiated highlands (H)

Source: Adapted from Köppen

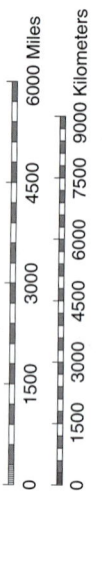

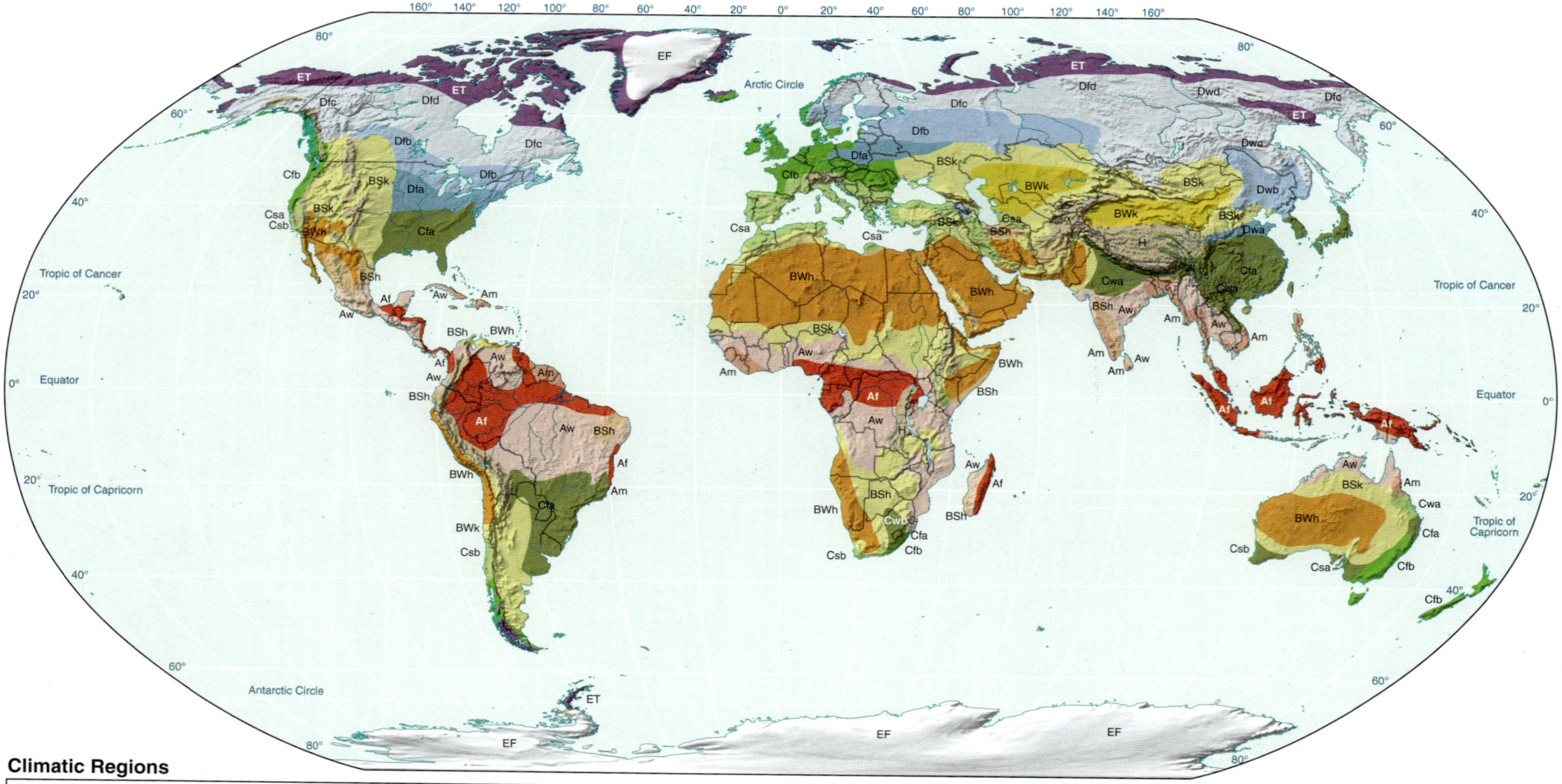

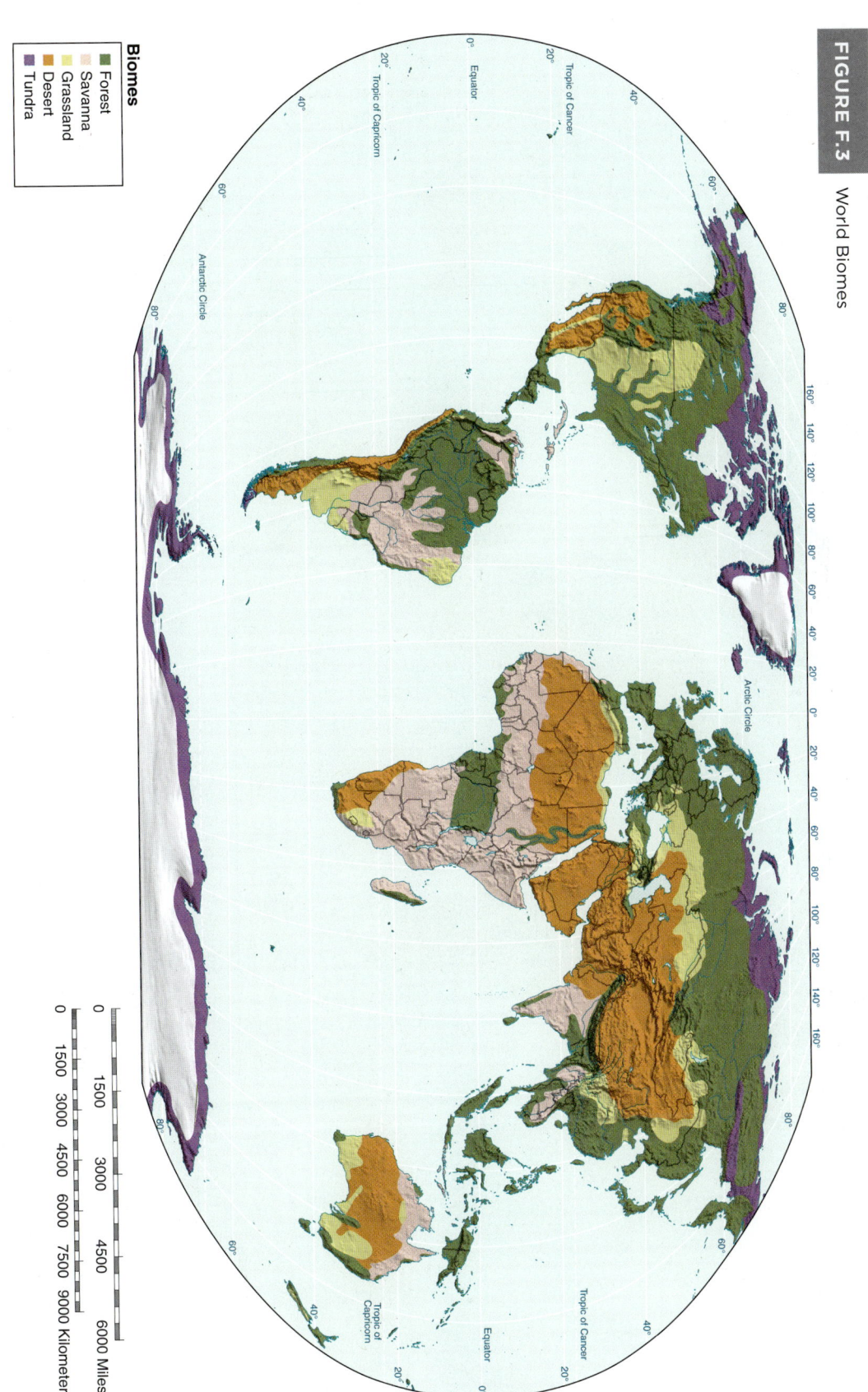

FIGURE F.3 World Biomes

FIGURE F.4 World soils

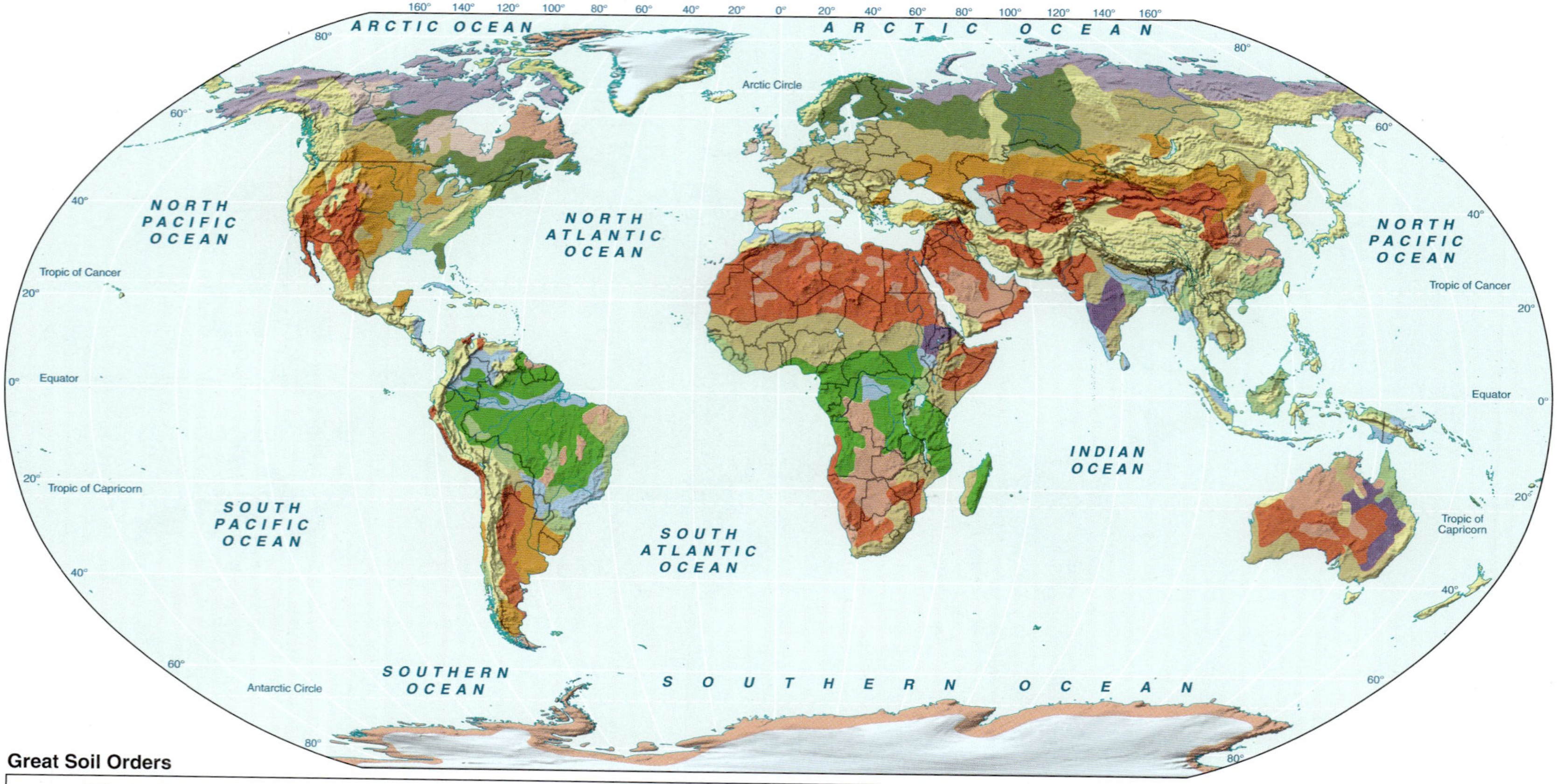

Great Soil Orders

- **Alfisols**: Grey to brown surface soils; medium to high base nutrients and organic content
- **Aridisols**: Dry or desert soils; high in base nutrients and low in organic content
- **Entisols**: Soils with poorly developed layers; typically wind-deposited soils
- **Histosols**: Swamps and bog soils; wet, highly organic (peat and muck) content
- **Gelisols**: Mineral or organic soil materials having permafrost within 100 cm of soils surface, with cryoturbation and/or ice segregation in the active layer
- **Inceptisols**: Weakly developed immature soils; typically tundra or volcanic soils
- **Mollisols**: Thick, dark soils of tallgrass praries; high in organic content and base nutrients
- **Oxisols**: Tropical and subtropical highly weathered soils; low in organic content and base nutrients
- **Spodosols**: Acidic soils of cool, moist forest regions; high organic content and low in base nutrients
- **Ultisols**: Acidic and clayey soils of upland tropical savannas; medium base nutrients
- **Vertisols**: Clay soils of moist tropical savannas; tend to crack and swell when dry
- **Mountain soils**: Thin soils, tending toward acidic; mixed varieties based on vertical zonation
- Little or no soil

Source: Adapted from Allen 2003

328

INDEX

ablation, 255
accumulation, in glacier formation, 255
actual evapotranspiration
    in decomposition rates, 126–27
    estimating net primary production and decomposition from, exercise, 129–30
    as soil moisture budget variable, 97
actual lapse rate, 49
adiabatic lapse rate, 49
adiabatic processes
    and orographic precipitation, exercise, 55–56
    overview, 49–50
adiabatic temperature change, 49
advancing glaciers, 256
aggradation, in drainage basins, 246
Agriculture, US Department of, 164
air masses, 59
    weather map symbols, 62
air pressure. *See* atmospheric pressure
air temperature. *See* atmospheric temperature
albedo, 12
alluvial terraces, 246–47
alluvium, 246
alpine glaciers, 255
    alpine glacial processes and landforms, exercise, 259–65
    landforms, overview, 256–57
anemometers, 33
apparent change in sea level, 274
applied climate classification system, 72
area, units of measure and conversions, 287
arêtes, 257
ash fall, 206
atmosphere, water in. *See* water, atmospheric
atmospheric path length, 1
atmospheric pressure, 31–33
    mapping, exercise, 35–37
    mapping and interpreting, overview, 31
    subpolar low-pressure cells, 71
    subtropical high-pressure cells, 71
atmospheric temperature
    continentality and (*see* continentality)
    introduction, 19–20

    latitudinal gradients, 20
    relationship between humidity and, 42–43
available water capacity, 113

backshore zone, 273
backwash, 271, 272
balance year, 255–56
bankfull discharge, 225
bar and swale topography, 246
barometers, 31
barren, of vegetation, 141
barrier islands, 273
bars, in atmospheric pressure, 31
basalt, 205–6
base level, of streams, 245
bathymetric contours, 186
baymouth bars, 273
benchmarks, in topographic maps, 186
Big Cyprus Swamp (Florida), 125, 126
bioclimatic communities.
    See also vegetation communities
    bioclimatic transects, exercise, 157–62
    bioclimatic transects data, overview, 152, 154–56
biomass, 125
biomes, 141, 164
    world map, 327
biotemperature in Holdridge's climatic variables, overview, 149–50
braided channels, 245
broadleaf foliage, 141
Buteo solitarius, 175–76

calderas, 206
capillary water, 113
carrying capacity, 175–76
Cascade Range, 206
Celsius temperature scale, 19
Cibotium glaucum, 175
cinder cone habitats, 176
cinder cones, 205, 206
cirques, 257
classification, as tool of science, 72
clay, in soil texture, 111, 112

329

climate classification, 71–85
  exercise, 77–85
  introduction, 71–75
  Köppen climate classification, 72, 74
  procedure for, 72–73, 75
  types of systems, 72, 73
climates
  classification (*see* climate classification)
  coincident (*see* coincident climates)
  global climate controls, 71–72
  regional, 71–72, 325
  in soil moisture budgets, 98
  types, 72, 73
  vegetation, climatic limits on (*see* vegetation communities)
  world map, 325
clouds, development, 50
coastal landforms, 271–85
  currents in, 271
  exercise, 275–85
  introduction, 271–74
  sea level, changes in, 274
  wind and waves in, 271
coastal zone, 271–73
coincident climates, 163–74
  climate and plant growth, variables for, 163–64
  data compilation and classification, exercise, 167–71
  index of moisture, 164
  interpretive applications, exercise, 173–74
  introduction, 163–64
  spatial patterns, 165
  vegetation and soils in, 164
cold fronts, 59
Colorado, climates and vegetation, 152, 154–55
composite volcanoes, 205, 206
concave surface, 223
condensation, 41–42
condensation level, 50
conduction, 11, 13
continental glaciers
  continental glacial landforms, exercise, 267–69
  overview, 255, 257–58
continentality, 19–20
  and air temperature, exercise, 23–25
  in soil moisture budgets, exercise, 107–10
continuous vegetation, 141
contour intervals, 186
contour lines, 186–89
convection, 11, 13
  in cloud development, 50
  instability and convectional rainfall, exercise, 57–58
convectional uplift, 59
convergent uplift, 59
convex surface, 223

cooling process, 41
Crater Lake National Park and Vicinity (Oregon) (topographic map), 308
Crater Lake (Oregon), 206
crevasses, 256
currents, ocean. *See* ocean currents
cut-banks, 246

damaged rainforest habitats, 175
day length period, 1, 4
decay time, 127
December solstice, 2
deciduous plants, 141
declination of the sun, 2
  exercise, 5–9
  in sun angle, 3–4
decomposition
  actual evapotranspiration in decomposition rates, overview, 126–27
  estimating net primary production and decomposition from actual evapotranspiration, exercise, 129–30
  mapping and map interpretation of, exercise, 131–40
deficits, in soil moisture budgets, 97
degradation, in drainage basins, 247
density, vegetation, 141
deposition, 224
depression contours, 186, 187
deserts, 142, 164
dew point, 43, 44, 50
diffuse shortwave radiation, 12
dikes, volcanic, 207
direct beam shortwave radiation, 12
dispersal, 141
dispersion, 141
disturbed rainforest habitats, 175–76
diversity, of forest biomes, 141–42
drainage basin analysis, 221–43
  characteristics, 223–24
  delimiting, 221–23
  drainage basin response, 224–25
  flood magnitude and frequency, exercise, 239–43
  floods and flood frequency, 225–26
  introduction, 221–26
  storm hydrographs, exercise, 235–37
  surface runoff and erosion processes, exercise, 231–34
  topography and drainage basins, exercise, 227–29
drainage divides, 221–23
droughts, 96
drumlins, 258
dry adiabatic lapse rate, 49
dry-bulb thermometer, 43
dry climates, 98
dunes, 273–74

ecosytems, 141–42
    damaged habitats, 175
    disturbed habitats, 175
    Hawaii Volcanoes National Park, ecosystem study (*see* Hawaii Volcanoes National Park)
ecotones, 175
effective moisture, 95
emergent coastlines, 274
empirical climate classification system, 72
end moraines, 257
energy
    balance (*see* energy balance)
    exchange between earth's surface and atmosphere, 11
    radiation balance (*see* radiation balance)
    solar radiation (*see* insolation)
    units of measure, 288
energy balance
    geographic variations in, exercise, 17–18
    overview, 13–14
energy fluxes, 13
environmental lapse rate (ELR), 49
Environmental Protection Agency (EPA), 125
equilibrium line, in glacier formation, 256
erosion
    overview, 224
    from streams, 245
    surface runoff and erosion processes, exercise, 231–34
eustatic changes in sea level, 274
evaporation, 41, 95–96, 163–64
evapotranspiration, 95, 163–64
    actual (*see* actual evapotranspiration)
evergreens, 141

Fahrenheit temperature scale, 19
fetch, 271
field capacity, 95, 113
firn, 255
firn lines, 255
floodplains
    definition, 225, 245–46
    in flood frequency, 225–26
    landforms, 245–47
floods and flood frequency
    flood magnitude and frequency, exercise, 239–43
    overview, 225–26
fluvial landforms, 245–53
    floodplain landforms, 245–47
    introduction, 245–47
    landforms of meandering streams, exercise, 251–53
    river terraces: Snake River (Wyoming), exercise, 249–50
    stream channel patterns, 245
foredunes, 273–74
foreshore zone, 271–73
forest biome, 164

forests, 141–42, 175
form, plant
    overview, 141
    range and form, exercise, 143–46
formation groups, 164
frequency, of flooding, 225–26
frontal uplift, 59
fronts, weather, 59

genetic climate classification system, 72
Geologic Survey, United States. *See* United States Geological Survey (USGS)
glacial flour, 257
glacial landforms, 255–69
    exercises, 259–69
    glacial mass balance, 255–56
    introduction, 255–58
glacials, 274
glaciers
    alpine (*see* alpine glaciers)
    continental (*see* continental glaciers)
    formation, 255
    movement, 256, 257
global energy balance, 11
gradients, in topographic maps, 187
gradients, latitudinal temperature, 20
graphic scale, 185–86
grasses, 141
grasslands, 142, 164
gravity water, 113
ground heat flux, 13
ground moraines, 257–58
gullies, 224

habit, 141
hanging valleys, 257
hapu'u pulu tree fern, 175
Hawaii
    rainforest recovery, 175
    volcanic activity, 175, 176, 205–6
Hawaiian hawk, 175–76
Hawaii'i o'o honeyeater, 175–76
Hawaii Volcanoes National Park
    assessing habitat changes, exercise, 181–84
    Devastation Trail study area, aerial photo, 302, 303
    Devastation Trail study area, overview, 175–76
    ecosystem study, 175–84
    feature and habitat identification, exercise, 179–80
    habitats, 175–77
    plant identification, 176–77
    scale conversion, 177–78
heat, 19
herbs, 141
hillslopes, 221–23

331

Holdridge, L.R., 149
Holdridge's ratio, 150, 151
horizons, soil, 111
horns (mountain peaks), 257
hot spots, 205
humidity, 41
    measuring, exercise, 47–48
    measuring, overview, 42, 43
    relationship between air temperature and, 42–43
hundred-year floods, 226
hydrographs
    overview, 225
    storm hydrographs, exercise, 235–37
hydrologic cycle, 223–24
hygroscopic water, 113
hypotheses, 113

ice ages, 255, 257
Iceland, volcanic activity, 205–6
igneous landforms, 205–20
    exercise, 209–20
    introduction, 205–7
    volcanic-landscape alteration, 206–7
incipient oxbow lakes, 246
incoming longwave radiation, 12
index contour lines, 186
index of moisture, 163, 164
infiltration capacity, 111
infiltration rate, 111, 224
insolation, 1
    composition, 12
    exercise, 5–9
    radiation balance (*see* radiation balance)
    sun angle, 1–4
interception, of water, 224
interlobate moraines, 258
intermediate contour lines, 186
intertropical convergence zone, 71
'Io hawk, 175–76
isobars, 31, 32, 60
isolated vegetation, 141
isolines, drawing, 290–91
isostatic changes in sea level, 274
isothermal layer, 49

June solstice, 2

kettle holes, 258
Kilauea Iki volcano (1959), 175, 176
Kilimanjaro, Mount (Africa), 206
Köppen, Vladimir, 72
Köppen climate classification, 72, 74
Küchler, August Wilhelm, 141

lag time, in drainage basins, 225
landforms
    fluvial (*see* fluvial landforms)
    glacial (*see* glacial landforms)
lapse rates
    overview, 49
    and stability, exercise, 51–54
large-scale maps, 186
latent heat, 41–42
latent heat flux, 13
latent heat of vaporization, 49–50
lateral erosion, 245
lateral moraines, 257
latitude
    in soil moisture budgets, exercise, 107–10
    in sun angle, 1–3
    temperature gradients (*see* latitudinal temperature gradients)
latitudinal temperature gradients
    in North America, exercise, 27–29
    overview, 20
lava, 176, 205–8
lehua tree, 175
length, units of measure and conversions, 287
lichen, 141
life zones, 150–51
Life Zone Triangle, 150
litter, 126–27
local relief, in topographic maps, 187
longshore currents, 271, 272
longwave radiation, 11, 12

magma, 205–7
magnitude, of flooding, 225–26
maps
    constructing profiles, 292–93
    isolines, 290–91
    topographic (*see* topographic maps)
    USGS (*see* United States Geological Survey (USGS))
map scale
    exercise, 201–4
    overview, 185–86
March equinox, 2
marine terraces, 273
mass balance, 255
Mather, John R. ("Role of Climate in the Distribution of Vegetation, The"), 165
Mauna Loa (Hawaii), 205
Mazama, Mount (Oregon), 206
meandering stream patterns
    landforms of meandering streams, exercise, 251–53
    overview, 245
meander scars, 246
measure, conversions and units of, 287–89

mechanical induction in air pressure, 31
medial moraines, 257
Metrosideros polymorpha, 175
millibars, 31
mixed broadleaf foliage, 141
mixing ratio, 42
Moho nobilis, 175
moisture. *See* water
moraines, 257–58
mountains, as climate controls, 71–72
mouth, of drainage basin, 221

National Park Service, 153
natural levees, 246
nearshore zone, 271
needleleaf foliage, 141
net primary production, 125–27
    actual evapotranspiration in, 125–26
    estimating NPP and decomposition from actual evapotranspiration, exercise, 129–30
    mapping and map interpretation of, exercise, 131–40
    production gradients, 126
net radiation, 12–13
New York, net primary production for Riverhead, 126, 127
nonforest biomes, 164
non-radiative fluxes, 13
noon sun angle
    exercise, 5–9
    overview, 1, 2
normal lapse rate, 49
Nunavut, life zone of Kugluktuk, 150–51

occluded fronts, 59
ocean currents
    as climate controls, 71
    in coastal landform formation, 271
    influence of currents on coastal climate, exercise, 87–94
'ohi'a lehua tree, 175
orographic precipitation
    adiabatic processes and, exercise, 55–56
    overview, 50
orographic uplift, 59
outgoing longwave radiation, 12
outgoing shortwave radiation, 12
outlet, of drainage basin, 221
outwash plains, 258
overland flow, 223–24
oxbow lakes, 246

parabolic dunes, 274
parabolic glacial valleys, 257
parklands, 142
patches of vegetation, 141

peak discharge, in drainage basins, 225
permeability
    in drainage basins, 224
    in soil moisture storage, 111, 113
pitted outwash plains, 258
plants. *See also* vegetation
    characteristics, 141
    communities (*see* vegetation communities)
    density (world map), 301
    foliage form (world map), 298
    foliages, 141, 298–301
    form, 141
    growth habit (world map), 300
    stature (world map), 299
plate tectonics, 205–6
plutonic igneous landforms, 205
pocket stereoscopes, using, 294–96
point bars, 246
polar front jet stream, 71
polar fronts, 59–60
pore spaces, in soil moisture storage, 111
porosity, in soil moisture storage, 111, 113
potential evapotranspiration, 95–96, 163–64
    in Holdridge's climatic variables, overview, 149–50
    as soil moisture budget variable, 96–97
prairies, 142
precipitation, 95, 149
    adiabatic processes in (*see* adiabatic processes)
    cloud development in, 50
    drainage basins (*see* drainage basin analysis)
    in Holdridge's climatic variables, overview, 149–50
    instability and convectional rainfall, exercise, 57–58
    lapse rates in (*see* lapse rates)
    overview, 49, 59–60, 163–64
    as soil moisture budget variable, 96–97
precipitation-potential evapotranspiration, as soil moisture budget variable, 96–97
pressure, units of measure and conversions, 288
pressure gradients
    mapping seasonal pressure gradients and wind speed, exercise, 39–40
    overview, 31, 33
pressure profiles, 31, 32
production gradients, 126
profiles, constructing map, 292–93
pyroclasts, 205

quadrats, in ecosystem studies, 176

radiation, 11, 13
radiation balance
    geographic variations in, exercise, 15–16
    longwave radiation, 11, 12
    net radiation, 12–13

radiation balance *(continued)*
    overview, 11–13
    shortwave radiation, 11–12
radiation intensity
    exercise, 5–9
    overview, 3
rainfall. *See* precipitation
rainforests
    characteristics, 175
    disturbance and succession, 175–76
    Hawaii Volcanoes National Park, ecosytem study of (*see* Hawaii Volcanoes National Park)
Rainier, Mount (Washington), 205
Rainier National Park, Mount (Washington) (topographic map), 316
range
    and dispersal, exercise, 147–48
    explained, 141
    and form, exercise, 143–46
ratios, in map scale, 185–86
recessional moraines, 257
recharge, in soil moisture storage, 97, 99
recurrence interval, 225–26
refraction, wave, 271, 272
regression, of sea level, 274
relative humidity, 42, 43, 45
representative fractions, 177–78, 185–86
retreating glaciers, 256
return period, 225–26
rills, 224
"Role of Climate in the Distribution of Vegetation, The," (Mather & Yoshioka, 1968), 165

sand
    in coastal landform formation, 273–74
    in soil texture, 111, 112
saturated adiabatic lapse rate (SALR), 50
saturation, 42–43
savannas, 142, 164
scale
    graphic scale, 185
    map scale (*see* map scale)
    in vertical-perspective aerial images, 177–78
scientific method, 113
sea cliffs, 273
sea level, changes in, 274
seasons, soil moisture, 99
sea stacks, 273
secondary dunes, 273–74
sediment, in coastal landform formation, 273–74
seed habits, 141
semideciduous plants, 141
sensible heat, 41–42
sensible heat flux, 13
September equinox, 2

shadow, in vegetation interpretation, 176, 177
shield volcanoes, 205–6
short herbs, 141
shortwave radiation, 11–12
shrubs, 141
silica, 205–6
sills, volcanic, 207
silt, in soil texture, 111, 112
sling psychrometer, 43
slope gradient, in surface runoff, 224
slope length, in surface runoff, 224
slopes, hill, 221–23
small-scale maps, 186
snag habitats, 176
Snake River (Wyoming), river terraces, exercise, 249–50
soil
    defined, 111
    great soil orders (world map), 164
    relationship of biomes and formation groups to, 164
soil moisture budgets, 95–110
    comparative field capacities, exercise, 105–6
    concept, 95–96
    example, 97–98
    graphs, 99
    introduction, 95–99
    latitude, altitude, and continentality, exercise, 107–10
    tables and graphs, exercise, 101–4
    variables, 96–97
soil moisture properties, 111–23
    analysis and interpretation, exercise, 117–22
    introduction, 111–13
    soil moisture analysis, exercise, 114–16
soil moisture seasons, 99
soil moisture storage
    porosity and permeability in, 111, 113
    as soil moisture budget variable, 97
soil orders, 164
soil texture, 95, 111, 112
soil texture class, 111, 112
soil texture triangle
    exercise, 123
    overview, 111, 112
solar constant, 1
solar radiation. *See* insolation
solstice, 2
specific heat, 19
specific humidity, 42
spits, 272, 273
St. Helens, Mount (Washington) (topographic map), 309
stable air, 50
statements, in map scale, 185–86
stationary fronts, 59
stationary glaciers, 256
stature, of vegetation, 141
steppes, 142

stereoscopes, using pocket, 294–96
storm hydrographs, exercise, 235–37
straight stream patterns, 245
stratovolcanoes, 205, 206
stream discharge, 224–25
streams
    base level, 245
    channel patterns, 245
    in drainage basins (*see* drainage basin analysis)
    floods and flood frequency, 225–26
    fluvial landforms (*see* fluvial landforms)
    yazoo, 246
submergent coastlines, 274
subpolar low-pressure cells, 71
subsurface rootstock, 141
subtropical high-pressure cells, 71
succession, in forests, 175
summer season, in glacier formation, 255–56
sun angle
    earth-sun geometry and, exercise, 5–9
    earth-sun geometry and, overview, 1–4
supplementary contour lines, 186
surface runoff
    overview, 223–24
    surface runoff and erosion processes, exercise, 231–34
surplus, in soil moisture budgets, 96, 97
swash, 271

tangent of an angle, 289
tarns, 257
temperature, 19
    air (*see* atmospheric temperature)
    conversions, exercise, 21
    influence on coincident climates, 163–64
    scales and statistics, 19
    units of measure, 288
temperature inversions, 49
tephra, 205
terminal moraines, 257
terrace scarps, 246, 247
terrestrial biomes, 141–42
texture, in vegetation interpretation, 176
thermal induction in air pressure, 31
throughflow, 224
tilt of earth's axis, 1, 2
tone, in vegetation interpretation, 176–77
topographic maps, 185–204
    drawing contour lines, exercise, 191
    introduction, 185–90
    map scale, exercise, 201–4
    map scale, overview, 185–86
    symbols, 190
    topographic map interpretation, exercise, 193–99
    topography, 186–89
    USGS (*see* United States Geological Survey (USGS))

topographic profiles, 188, 189
total relief, in topographic maps, 187
transect
    bioclimatic transect data, 152, 154–56
    defined, 152
transgression, of sea level, 274
transpiration, 95–96, 163–64
trees, 141
troughs, 59
tundra, 142, 164

United States Geological Survey (USGS)
    benchmarks, placement of, 190
    Chief Mountain (Montana, Alberta), 306
    Compti (Louisiana) (topographic map), 315
    Crater Lake National Park and Vicinity (Oregon), 308
    Folsom (New Mexico) (topographic map), 310
    forests defined, 141
    Hartford (Alabama) (topographic map), 305
    Moose (Wyoming) (topographic map), 314
    Mount Rainier National Park (Washington) (topographic map), 316
    Mount St. Helens (Washington) (topographic map), 309
    Portions of Bayshore East and Sayville (New York) (topographic map), 320
    Portions of Provincetown and North Truro (Massachusetts) (topographic map), 319
    Redondo Beach (California), 321
    Rome (Wisconsin) (topographic map), 318
    Saint Mary (Montana, Alberta), 307
    Ship Rock (New Mexico) (topographic map), 312
    standardization of topographic map symbols, 190
    Thorofarre Buttes (Wyoming) (topographic map), 313
    Thousand Springs (Idaho) (topographic map), 304
    Whitewater (Wisconsin) (topographic map), 317
unstable air, 50
upwelling of ocean water, 71
U-shaped glacial valleys, 257
utilization, in soil moisture budgets, 97, 99

vapor pressure, 42
vegetation
    characteristics, 141
    classification, 142–43
    in coastal landform formation, 273–74
    in coincident climates, 163–65
    communities (*see* vegetation communities)
    foliage form (world map), 298
    foliages, 141, 298–301
    growth habit (world map), 300
    Hawaii Volcanoes National Park, ecosystem study (*see* Hawaii Volcanoes National Park)
    plant density (world map), 301
    plant stature (world map), 299
    range and dispersal, exercise, 147–48

vegetation *(continued)*
    range and form, exercise, 143–46
    terrestrial biomes, 141–42
vegetation communities, 149–62
    bioclimatic transect data, overview, 152, 154–55
    bioclimatic transects, exercise, 157–62
    climatic limits on, introduction, 149–56
    confirming vegetation predictions, overview, 151, 153
    Holdridge's climatic variables, overview, 149–50
    predicting, overview, 150–51
velocity, units of measure and conversions, 288
vertical erosion, 245
vertical exaggeration, in topographic maps, 188, 189
volcanic necks, 207
volcanoes, 205–7
    definition, 205
    igneous landforms from, 205
    landscape alteration by, 206–7
    types, 205–6
volume, units of measure and conversions, 287
V-shaped glacial valleys, 257

warm fronts, 59
water
    capillary, 113
    in coincident climates, 163–65
    effective moisture, 95
    gravity, 113
    index of moisture, 163, 164
    soil moisture budgets *(see* soil moisture budgets)
water, atmospheric, 41–48
    humidity *(see* humidity)
    introduction, 41–43
    moisture phase changes, exercise, 47–48
    moisture phase changes, overview, 41–42
water budget tables, 95
wave-cut notches, 273

wave-cut platforms, 273
waves, in coastal landform formation, 271
weather
    atmospheric elements in, 59–62
    fronts, 59
    precipitation *(see* precipitation)
weather maps
    interpretation, exercise, 63–69
    reading and interpreting, introduction, 31–33, 60–62
    symbols, 61–62
weather stations, 33, 60
weight, units of measure and conversions, 288
wet-bulb depression, 43
wet-bulb thermometer, 43
wet climates, 98
wilting point, 113
wind
    in coastal landform formation, 271, 273–74
    direction, 31, 33
    speed *(see* wind speed)
wind speed
    mapping seasonal pressure gradients and, exercise, 39–40
    overview, 31, 33, 60
    symbolization, 33, 61
winter season, in glacier formation, 255–56
Wisconsin advance, 257
woodlands, 142
world physiography (world map), 325

yazoo streams, 246
Yoshioka, Gary A. ("Role of Climate in the Distribution of Vegetation, The"), 165

zenith angle
    exercise, 5–9
    overview, 1, 2